有机化学实验

（英汉双语版）

主编　张大伟

科学出版社

北　京

内 容 简 介

本书是编者根据多年教学经验，结合农林院校本科非化学专业的需求和特点，汲取高等农林院校教学改革和科学研究成果中的精华编写而成的。全书分为有机化学实验基础知识、有机化学实验基本操作、有机化学综合实验3章，涵盖了农学类专业本科有机化学实验教学大纲要求的内容。本书突出加强对学生中英文化学基础知识的掌握和训练，注重学生化学专业英语水平的提高，为其后续深造奠定基础。

本书可作为高等农林院校农学类、食品类、动物医学类、生物技术类各专业本科生、研究生教学用书，也可作为化学相关专业教师和学生参考用书。

图书在版编目（CIP）数据

有机化学实验：英汉双语版 / 张大伟主编. —北京：科学出版社，2019.8

ISBN 978-7-03-057956-0

Ⅰ. ①有… Ⅱ. ①张… Ⅲ. ①有机化学-化学实验-高等学校-教材-英、汉 Ⅳ. ①O62-33

中国版本图书馆CIP数据核字（2018）第131175号

责任编辑：陈雅娴 高 微 / 责任校对：樊雅琼
责任印制：张 伟 / 封面设计：迷底书装

科学出版社 出版
北京东黄城根北街16号
邮政编码：100717
http：//www.sciencep.com

北京中石油彩色印刷有限责任公司 印刷
科学出版社发行 各地新华书店经销
*
2019年8月第 一 版 开本：720×1000 B5
2022年8月第二次印刷 印张：10
字数：202 000

定价：38.00元

（如有印装质量问题，我社负责调换）

《有机化学实验（英汉双语版）》编写委员会

主　编　张大伟（吉林大学）

副主编　申贵男（黑龙江八一农垦大学）

　　　　李秀花（吉林农业科技学院）

　　　　卢　可（吉林大学）

　　　　邹　楠（吉林大学）

　　　　邹连春（吉林大学）

　　　　贾赞慧（吉林大学）

　　　　邵奎占（东北师范大学）

　　　　刘　伟（沈阳农业大学）

参　编　杨　桦（长春工程学院）

　　　　许迪欧（吉林大学）

主　审　徐显秀（山东师范大学）

　　　　李　婧（黑龙江八一农垦大学）

前　　言

有机化学实验是高等农林院校非化学专业重要的基础必修课程，是学生获得全面化学素质教育的一个重要环节，对农科学生掌握化学基础知识，培养实践能力、创新精神等起着不可替代的作用。尤其近些年，大学生素质日益提高，各高等学校针对优秀学生实行卓越、拔尖计划，单独建班进行培优教育；同时，学生对外交流机会日益增多，参与创新实验训练热情高涨，因此学生在有机化学实验学习中迫切需要提高其化学专业英语水平和应用能力。为此，编者基于多年一线实验教学经验和体会，结合农林院校本科非化学专业的需求及特点，汲取高等农林院校教学改革和科学研究成果中的精华编写本书。

本书具备三大特色：注重基础，循序渐进；拓展交叉，创新培养；信息融合，多元教学。

（1）注重基础，循序渐进。全书分为有机化学实验基础知识、有机化学实验基础操作、有机化学综合实验 3 章，涵盖了农学类本科生有机化学实验大纲要求的内容，可满足当前各农林院校非化学专业的普遍需求。在本书编写过程中，从实验基础操作到综合实验，都突出加强学生对中英文化学基础知识的掌握和训练，为学生后续学习奠定坚实基础。

（2）拓展交叉，创新培养。融入双语教学内容，不仅有助于提高学生的化学专业英语水平，更是扩大交流、拓宽视野、满足个人未来发展需要的重要举措，同时也为学生尽早参加大学生创新实验、参与科学研究、培养个人创新能力提供了机会，弥补传统教学的短板。

（3）信息融合，多元教学。随着信息技术的快速发展，学生自主学习、网络学习已成为趋势，化学实验的教与学也已不局限于实验室中。本书配有微信教学公众号，可扫描封底二维码关注，公众号平台提供丰富的实验图片、动画、视频、音频等新媒体资源辅助教与学，便于学生碎片化学习，延展学生学习的空间和时间，实现“课前—课中—课后”三段立体混合式教学新体系的构建。

本书由吉林大学张大伟教授主编；副主编有黑龙江八一农垦大学申贵男，吉林农业科技学院李秀花，吉林大学卢可、邹楠、邹连春、贾赞慧，东北师范大学邵奎占，沈阳农业大学刘伟；参加编写的还有长春工程学院杨桦，吉林大学许迪欧；主审有山东师范大学徐显秀，黑龙江八一农垦大学李婧。全书由张大伟完成统稿、定稿等工作。

本书的编写得到吉林大学教务处和科学出版社的鼎力支持，并获吉林大学“十三五”规划教材立项支持，在此谨致以衷心的感谢。在编写过程中，编者参考了部分已出版的资料，在此对原作者一并致谢。

编者虽竭力做到全书内容的科学与准确，但限于水平，书中可能存在疏漏和不妥之处，恳请读者批评指正。

编　者

2019 年 3 月

目　　录

第1章　有机化学实验基础知识

化学是一门实验科学。在有机化学实验课程中，学生将在分子层面对化学反应获得基本认识，通过实验现象和结果理解为什么某些分子间化学反应需要在特定条件才能进行，同时，将有机会在这门实验课程中进行探索和发现，了解有机化学家常用的科学探究方法，思考理论与实验之间的关系，解释实验结果，加深对有机化学理论课堂上所学知识的理解。所以，有机化学实验课的主要目的是帮助学生了解有机化学实验的实际操作过程，学习如何获得和解释实验结果，并从中得出合理的结论。

通过有机化学实验课程可获得一些重要实验技能，如仔细观察和记录数据，与他人的团队合作，在文献中查找化学药品和化学反应信息的能力等，为后续化学实验和其他专业实验室课程的学习打下扎实实验基础，同时培养辩证思维能力，养成严谨的科学态度和工作习惯。

1.1　有机化学实验基本规则

在正式开始有机化学实验课前，每名学生必须了解和掌握化学实验室基本规则，下面所列各项要求必须牢记。

1）实验课前

（1）认真预习实验内容，明确实验目的要求，熟悉方法步骤，掌握基本原理，写好预习报告。

（2）认真清点实验仪器和药品，如有缺少、损坏应立即报告实验指导教师。

（3）实验室内必须穿实验服，佩戴护目镜及橡胶手套。

2）实验课中

（1）认真听实验指导教师讲解实验目的、步骤、仪器和药品性能、操作方法和注意事项。

（2）实验室内不得打闹、吸烟、吃东西。

（3）严格执行操作规程，仔细观察实验现象，认真做好实验记录，根据实验过程分析实验结果，写出实验报告。

（4）注意安全，使用腐蚀性强、易燃、易爆和有毒药品要小心谨慎，如果在实验中发生意外事故不要惊慌，应立即报告实验指导教师处理。

（5）爱护仪器和实验材料，节约用水和实验材料。未经实验指导教师许可，不得将实验仪器和材料带出化学实验室。

3）实验课后

（1）实验完毕，在实验指导教师和实验准备教师的指导下清点好实验器材，归还原位，妥善处理废物并做好仪器清洗，保持桌面及室内整洁，经指导教师许可方能离开实验室。

（2）离开化学实验室之前，应检查水、电、门、窗是否关好，易燃、腐蚀性和有毒药品是否已妥善保管好。

1.2 有机化学实验安全常识

有机化学实验室是实验事故常发地。在实验室工作时，正确的实验操作能够避免大多数危险，不规范、不正确的实验操作可能造成严重的伤害。因此，实验安全是关乎每个人的重要大事。

实验室事故常见类型有：火灾和爆炸、电器安全、烫伤、割伤或化学灼伤，以及因吸入、皮肤吸收或进食有毒物质而发生的事故。为减少实验室事故发生的可能性，应认真采取以下预防和处理措施。

1.2.1 有机化学实验室事故的预防和处理

1. 火灾

有机化学实验室中使用的有机溶剂大多数易燃，火灾是实验室常见事故之一，应尽可能避免使用明火，同时还应注意勿将易燃液体放在敞口容器（如烧杯）中明火直接加热。当附近有露置的易燃溶剂时，切勿点火。实验室内易燃易爆溶剂储存应适量，并放置在规定区域。当处理大量的可燃性液体时，应在通风橱中或在指定地方进行，室内应无火源。经常检查气阀和煤气灯，使其处于良好状态，防止泄漏。

在火灾事故发生时要保持冷静、及时处理，通常采取以下措施：

（1）防止火势蔓延。应立即扑灭火灾源头，切断电源，清除易燃物。

（2）根据火灾程度，采取不同方式立即扑灭。如火灾区域小，用防火棉或沙子覆盖；如火灾区域大，使用灭火器。如扑灭有机物火灾，不要用水；如发生电器着火，首先要切断电源，然后用干冰灭火器灭火。

总之，当失火时，应根据起火的原因和火场周围的情况，采取不同的方法灭火。无论使用哪一种灭火器材，都应从火源的四周开始向中心扑灭，把灭火器的喷口对准火焰的底部，在灭火过程中切勿犹豫。常用灭火器类型和特点见表1.1。

表 1.1　常用灭火器名称及用途

中文名称	型号简写	适用范围
泡沫灭火器	MP	适用于扑灭油类着火，不能用于扑灭电器着火
二氧化碳灭火器	MT	适用于扑灭电器设备着火和小范围内油类及忌水化学品着火
干粉灭火器	MF	适用于扑灭电器设备、油类、可燃气体、精密仪器、图书资料及忌水化学品着火

2. 爆炸

有机化学实验预防爆炸的措施如下：

（1）装置必须正确安装，不能形成密闭体系，应使装置与大气相通，否则易发生爆炸。

（2）切勿使易燃易爆的气体接近火源，否则可能会由一个热的表面或者一个火花、电火花而引起爆炸。

（3）使用乙醚等醚类试剂时，必须检查有无过氧化物存在，如果发现有过氧化物存在，应立即用硫酸亚铁除去过氧化物才能使用；同时使用醚类试剂时，应在通风较好的地方或在通风橱内进行。

（4）对于易爆炸的固体，如重金属炔化物、苦味酸金属盐、三硝基甲苯等都不能重压或撞击，以免引起爆炸。这些残渣危险，必须小心销毁。

3. 电器安全

实验室内使用电器时，应防止人体与电器导电部分直接接触，不能用湿手或用手握湿的物体接触电源插头。为了防止触电，应仔细检查装置插头是否正确安装，电线是否松动，正确连接地线，以确保其处于良好状态。实验后应关闭电源，再将电源插头拔下。

4. 烫伤和割伤

了解急救箱的位置及箱内物品，以应急治疗简单的烫伤和割伤。

加热玻璃器皿时，不要用手直接触摸。不要把热玻璃器皿放在他人可能使用的椅子上。蒸汽和沸水会造成严重烫伤，小心处理装有沸水的容器。烫伤轻者可以用冷水浸泡 10～15 min，涂以烫伤药膏；重伤者应立即送医院诊治。

割伤是实验室内常见事故，受伤后要仔细观察伤口有没有玻璃碎粒，如有，应先把伤口处的玻璃碎粒取出。若伤势不重，先进行简单的急救处理，再用纱布包扎；若伤口严重，应按压止血，并立即送急诊室或医院治疗。

5. 化学药品灼伤

如果有化学药品溅到皮肤上，首先要用清水冲洗 10～15 min（另有特别说明药品除外）。这种处理方法会冲洗掉多余的化学试剂。对于酸、碱和有毒化学物质，用水彻底清洗后可以减轻疼痛。与强碱接触的皮肤通常不会立即产生疼痛或刺激，但如果不立即用大量的水冲洗受伤区域，可能会发生严重的组织损伤（尤其是眼睛）。如果化学灼伤较轻，可使用烧伤膏处理；严重的化学灼伤应立即就医治疗。

如果有化学物质进入眼睛，应立即去洗眼器处，用大量微温水清洗，冲洗眼球 10～15 min，使用洗眼液后立即就医。

6. 中毒

（1）不要直接用鼻子闻试剂。

（2）不得在实验室内品尝或喝任何试剂。

（3）实验后的有毒残渣必须做妥善而有效的处理，不能乱丢。

（4）有些毒性物质会渗入皮肤，因此接触这些物质时必须戴橡胶手套，操作后应立即洗手，切勿让毒物接触五官或伤口。例如，氰化钠接触伤口后就会随血液循环至全身，严重的会造成中毒死伤事故。

（5）在反应过程中可能生成有毒或有腐蚀性气体的实验应在通风橱内进行，使用后的器皿应及时清洗。在使用通风橱时，实验开始后不要把头部伸入橱内。

（6）如出现中毒症状，应及时进行治疗。如果有毒化学物质溅到嘴里，立即吐出，用大量的水漱口。如果已经吞食，应根据中毒性质进行急救，具体如下：①腐蚀性酸性毒物，先饮大量的水，再服用氢氧化铝膏、鸡蛋白或牛奶；②腐蚀性碱性毒物，先饮大量的水，然后服用醋、酸果汁、鸡蛋白或牛奶，不论酸或碱中毒都不要吃呕吐剂；③刺激性及神经性中毒物，先服牛奶或鸡蛋白使其缓和，再服用硫酸铜溶液（约 30 g 溶于一杯水中）催吐，有时也可以用手指伸入喉部催吐后，立即到医院就诊；④吸入气体中毒，将中毒者移至室外，解开衣领及纽扣，吸入少量氯气或溴蒸气者，可用碳酸氢钠溶液漱口。

1.2.2 化学实验废弃物处置

正确处理化学实验废弃物是有机化学实验课程学习中的重要内容之一。对化学实验废弃物的妥善处理，可以最大限度减少对环境的危害，减少学校处理化学实验中必须产生的废物的经济成本。因此，每名学生都必须增强环保意识，规范处理各类化学废弃物。

（1）废溶剂应放入废液桶中，并贴上标签。应注意避免不加区分地倒入混合溶剂，特别是卤化溶剂应与其他溶剂分开，避免发生反应。

（2）实验中使用的一些无害材料可以放入普通的废物桶中，如无害的固体废物、普通的纸或布等。

（3）破碎的玻璃器皿或滴管等玻璃制品应放入废玻璃桶中。

（4）有毒固体应密封并放置在单独的回收容器中。

（5）所有废物箱都应贴上标签，清楚标明内容物。

（6）所有废物容器在不使用时应加盖封闭。

1.3　有机化学实验常用玻璃仪器和实验装置

1.3.1　普通玻璃仪器

有机化学实验常用玻璃仪器如图 1.1 所示。

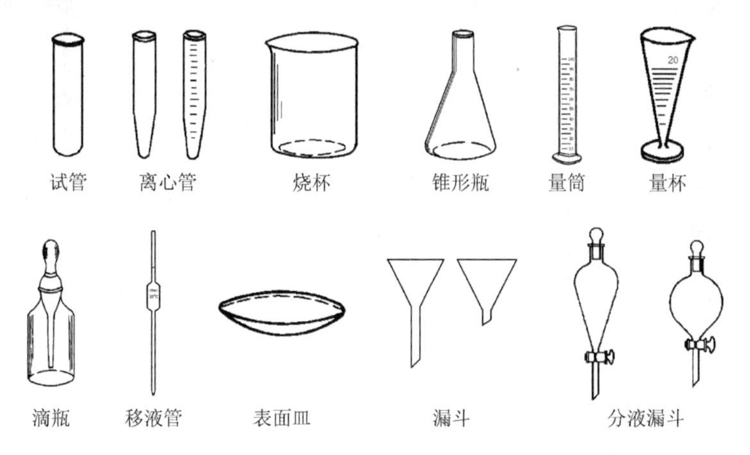

图 1.1　常用玻璃仪器

1.3.2　标准磨口仪器

在有机化学实验中，通常使用由硬质玻璃制成的标准磨口仪器。由于玻璃仪器的大小及用途不同，标准磨口的大小也不同，通常使用的标准磨口仪器常用两个数字表示磨口大小，数字间用斜线分隔，如 10/30，表示此锥形磨口端直径 10 mm，磨口长度为 30 mm。凡属同类规格的标准内外磨口均可互相紧密连接，因此可根据需要选配和组装各种类型的成套仪器。

常用标准磨口仪器如图 1.2 所示。

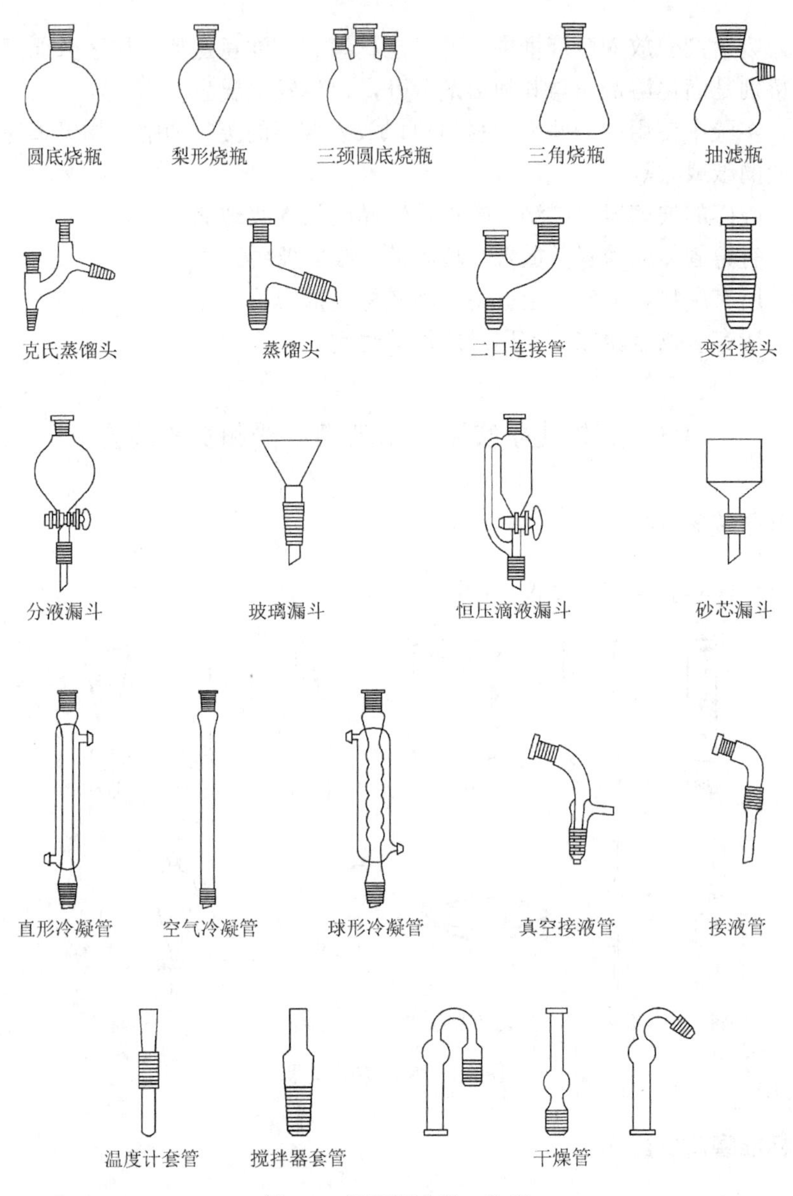

图 1.2　常用标准磨口仪器

1.3.3　常用成套实验装置

有机化学实验装置装配是否正确与实验成败有很大关系。

首先，在装配一套装置时，所选用的玻璃仪器和配件都要洁净，否则会影响产物的产量和质量。其次，所选用的器材要恰当。例如，在需要加热的实验中，如需选用圆底烧瓶时，其容积大小应为所盛反应物占其容积 1/2 左右为好，最多不应超过 2/3。最后，装配时应根据热源选准反应器的位置，并用铁夹固定反应器，然后按照先下后上、从左至右的顺序逐个装配起来，在拆卸时按相反的顺序逐个拆卸。

仪器装配要求做到严密、正确、整齐和稳妥。安装正确的实验装置既要合理实用，又要注意整齐美观。

常用成套实验装置如图 1.3～图 1.5 所示。

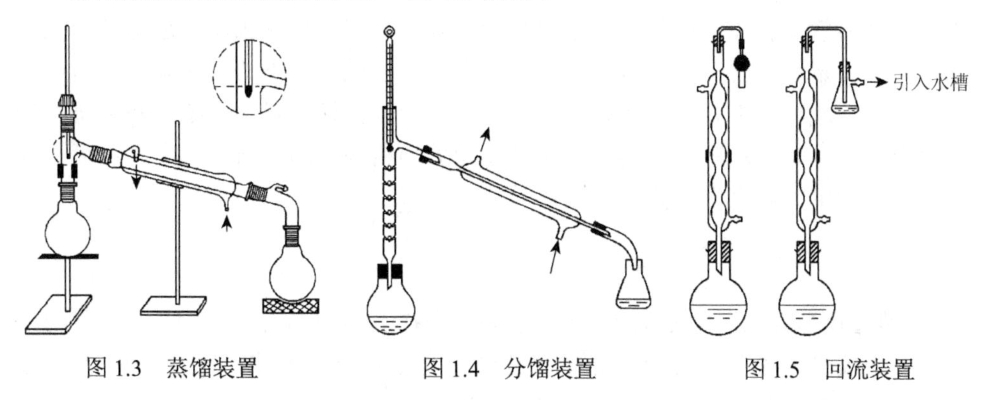

图 1.3　蒸馏装置　　图 1.4　分馏装置　　图 1.5　回流装置

1.4　有机化学文献

1.4.1　工具书

（1）*CRC Handbook of Chemistry and Physics*：年卷，美国（CRC Press，Boca Raton，FL）出版的一本化学与物理英文手册。它包含了化学和物理学所有领域大量有用数据。对有机化学最有用的部分为“有机化合物的物理常数”，提供了大约 11000 个化合物的反应式、结构式、摩尔质量（相对分子质量）、密度、折射率、溶解度、颜色、熔点和沸点。

（2）*The Merck Index of Chemicals and Drugs*（第十四版）：美国（Merck Co.，Rahway，NJ）出版的一本辞典。该书简述了 10000 多种化合物的物理、化学和药理学性质，包括药物、有机化学品和试剂、无机物、农药和天然产物。

（3）*Beilstein Handbook of Organic Chemistry*：该书内容较全，是最大的有机化合物信息汇编。该书早期以德语出版，自 1984 后以英文出版。该书包含约 1000 万种有机化合物结构、制备和性质数据。

1.4.2　期刊

1. 中文期刊

《化学学报》（月刊）创刊于 1933 年，原名《中国化学会会志》（*Journal of the*

Chinese Chemical Society），是我国创刊最早的化学学术期刊，1952 年更名为《化学学报》，并从英文版改成中文版。《化学学报》刊载化学各学科领域基础研究和应用基础研究的原始性、首创性成果，涉及物理化学、无机化学、有机化学、分析化学和高分子化学等。《化学学报》历史悠久，在我国化学界一直享有较高的声誉和学术地位。

《化学进展》（月刊）是由中国科学院基础科学局、化学部、文献情报中心和国家自然科学基金委员会化学科学部共同主办，以刊登化学领域综述与评论性文章为主的学术性期刊。读者可从中了解化学专业领域国内外研究动向、最新研究成果及发展趋势。主要栏目有：综述与评论，专题论坛，科学基金，基础研究论文评介，动态与信息等。该刊可供化学及相关学科领域的科研、教学、决策管理人员及大学生、研究生阅读。

2. 英文期刊

《美国化学会志》（*Journal of the American Chemical Society*，JACS），周刊。化学领域顶级期刊之一，JACS 由美国化学学会创办于 1879 年，已有 140 年历史。期刊收录既有简讯类文章，又有全文型文章，每年发表约 19000 页论著、通讯和述评。

《德国应用化学》（*Angewandte Chemie International Edition*，Angew. Chem. Int. Ed.），周刊。化学领域顶级期刊之一，由 Wiley 公司出版，分德语版和英语版。收录文章以简讯类为主，主要分布在有机化学、生命有机化学、材料学、高分子化学等领域。

《化学评论》（*Chemical Reviews*，Chem. Rev.），双周刊。一本同行评审的科学杂志，由美国化学学会于 1924 年创刊并发行至今。该刊发表某一领域内综合性评论，不发表原创实验研究。包括主题刊和混合刊两种期号；主题刊每期关注一系列主题，每个主题下有若干篇相关评论。

1.4.3 化学文摘

《化学文摘》（*Chemical Abstracts*，CA）创刊于 1907 年，是美国化学学会出版的世界上最全化学摘要期刊。每年 CA 将约 16000 种专业期刊、专利、评论、技术报告、专著、会议记录、专题讨论会和书籍，超过 100 万份文件的内容浓缩成摘要，并根据研究主题、作者姓名、化学物质或结构、分子式以及专利号对摘要进行索引。每个化合物都被分配一个编号，称为摘要号，它可以帮助读者很容易找到检索化合物的引用。

目前，CA 已普及电子化和网络化出版，并提供许多数据库，其中最具代表性、应用最广泛的数据库为 SciFinder 搜索引擎。

1.4.4　网上资源

随着网络技术的快速发展，从互联网上在线查找化学相关文献和资料等已变得非常方便、迅速，并已成为研究者必备的学习方法和手段。这里只对有机化学网络相关资源作简要介绍。

（1）网上图书馆。例如：

中国国家数字图书馆 http://www.nlc.cn/；

吉林大学图书馆 http://lib.jlu.edu.cn/portal/index.aspx/。

（2）期刊资源。例如：

中国知网 http://www.cnki.net/；

化学学报 http://sioc-journal.cn/Jwk_hxxb/CN/volumn/current.shtml；

化学进展 http://manu56.magtech.com.cn/progchem/CN/volumn/home.shtml。

（3）专利文献。例如，中国及多国专利审查信息查询 http://cpquery.sipo.gov.cn/。

（4）数据库资源。例如：

Chem Blink 化学数据库 https://www.chemblink.com/indexC.htm；

化学专业数据库 http://www.organchem.csdb.cn/。

1.5　有机化学实验预习和报告

1.5.1　预习报告

每名学生都需准备一本实验报告本，且必须在实验开始前仔细和全面地预习每个实验，这是化学实验学习的重要组成部分，因为仔细观察和准确的实验报告是科学实践的精髓。

预习报告的具体要求如下：

（1）预习教材中每个实验的说明及所要求的内容。

（2）理解反应原理（主反应和副反应）。

（3）写明主要试剂和产物的物理常数，主要试剂的规格和数量（g，mL，mol）；列出所需玻璃器皿。

（4）列出实验步骤和可能出现的化学危险。

（5）列出记录数据的表格。

1.5.2　实验记录

书写实验记录的指导原则是记录所有的细节，这些细节将使他人能够通过该实验记录理解所做实验，并且在事先不知情的情况下准确地重复整个实验。因此，在实验过程中，学生应规范操作，仔细观察，积极思考，并及时记录实验现象（如加热条件、颜色变化、酸碱度变化、沉淀生成、气体产生等）和实验数据（如室温、反应温度等）。重要的是将产量、熔点和沸点等结果记录在笔记本中。所有实验记录必须在实验时完成，绝不允许通过课后的记忆来补写笔记和记录结果。实验记录应真实、简洁、清晰。实验结束后，应向实验指导教师展示原始记录和获得的产品，并在离开实验室前进行检查。

对于有机合成实验，产率的高低和产品质量的好坏通常是评价一个实验结果及考核学生实验技能的重要指标之一。理论产量是假定反应完全转化成产物的质量，该数据可由反应方程式计算得出。实际产量是指实验中实际得到的产物质量。产率计算公式如下：

$$\text{产率} = \frac{\text{实际产量}}{\text{理论产量}} \times 100\%$$

1.5.3 实验报告

实验报告是学生实验学习的重要组成部分。一份完整的实验报告应包括实验目的（要做什么）、实验原理（为什么做）、实验试剂和仪器（如何做）和实验结果。实验结果的总结和结论尤其重要，还需要讨论实验失败的改进建议或解决方案。实验报告书写格式样例如下：

实 验 题 目

（1）实验目的（介绍实验中所使用的主要技术）。

（2）实验原理（化学反应，化学反应方程式或反应机理）。

（3）主要试剂和产品的物理常数。

（4）主要试剂和实验装置的规格和用量。

（5）实验装置图。

（6）实验步骤和实验现象（颜色变化，溶解性，沉淀等）。

（7）实验结论（产率，熔点，沸点等）。

（8）实验讨论。

样例：

正溴丁烷的合成

1. 实验目的

（1）学习从正丁醇制备正溴丁烷的原理及方法。

（2）掌握带气体吸收装置的回流、蒸馏及萃取基本操作。

2. 实验原理

主反应式：

$$NaBr + H_2SO_4 \longrightarrow HBr + NaHSO_4$$

$$CH_3CH_2CH_2CH_2OH + HBr \xrightarrow{H_2SO_4} CH_3CH_2CH_2CH_2Br + H_2O$$

3. 主要试剂及产品的物理常数

名称	相对分子质量	性状	相对密度 d_4^{20}	熔点/℃	沸点/℃	溶解度/(g/100 mL)		
						水	醇	醚
正丁醇	74.12	无色透明液体	0.8098	−88.9	117.25	7.9*	∞	∞
正溴丁烷	137.03	无色透明液体	1.2758	−112.4	101.6	不溶	∞	∞

* 20℃室温

4. 主要试剂的规格和用量

正丁醇（C.P.）18.5 mL（0.20 mol）；溴化钠（C.P.）25 g（0.24 mol）；浓硫酸（L.R.）29 mL（0.54 mol）；饱和 $NaHCO_3$ 水溶液；无水氯化钙（C.P.）2 g（0.54 mol）。

5. 实验装置

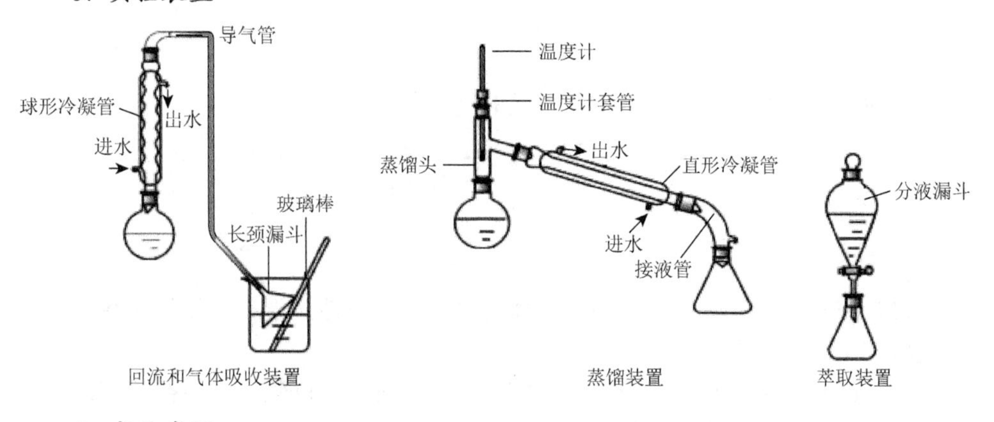

回流和气体吸收装置　　蒸馏装置　　萃取装置

6. 实验步骤

	步骤	现象
回流	向 150 mL 圆底烧瓶中加入 20 mL 水、29 mL 浓 H_2SO_4，振摇冷却。冷却至室温后，加入 18.5 mL n-C_4H_9OH 和 25 g NaBr，加一粒磁力搅拌子，摇匀。在圆底烧瓶瓶口安装冷凝管，冷凝管顶部安装气体吸收装置，开启冷凝水，磁力搅拌，电加热套加热回流 60 min。稍冷，拆除回流装置	放热； 反应中液体分层； 上层为正溴丁烷粗产物，颜色由淡黄色变为橙黄色

续表

	步骤	现象
蒸馏	拆除球形冷凝管和气体吸收装置，改为蒸馏装置，用 50 mL 锥形瓶收集馏分。蒸出正溴丁烷，直到馏出液变清（说明正溴丁烷全部蒸出）	开始馏出液为乳白色油状物，后来油状物减少
萃取	将蒸馏收集液倒入 50 mL 分液漏斗中，加入等量的水冲洗混合物，萃取、分液。收集下层正溴丁烷液体放入另一个干燥分液漏斗，在第二个漏斗中加入等量的浓硫酸，萃取、分液。弃去下部硫酸层。再次用 15 mL 水萃取有机相，然后用 15 mL 饱和碳酸氢钠水溶液萃取。注意：在最后一次萃取时，上水相应尽可能与下有机相完全分离	分层； 与水相比较，有机相在下层；与硫酸相比较，有机相在上层； 粗产物呈棕黄色
干燥	将粗产物转入干燥小锥形瓶中，加 2 g 无水氯化钙干燥 20 min，至液体澄清	开始浑浊，最后变清
蒸馏	将干燥后粗产物转入干燥的圆底烧瓶中，干燥剂应留在锥形瓶中。加入磁力搅拌子，蒸馏粗溴化正丁烷。收集 99～103℃馏分	温度很快升至 99℃，并稳定于 101～102℃，最后升至 103℃
产率	称量产物，计算产率	无色透明液体

7. 产率

实际产量：18 g

理论产量：其他试剂过量，理论产量按正丁醇计。

$$CH_3CH_2CH_2CH_2OH + HBr \xrightarrow{H_2SO_4} CH_3CH_2CH_2CH_2Br + H_2O$$

$$\begin{matrix} 1 & \qquad\qquad & 1 \\ 0.2 & \qquad\qquad & 0.2 \end{matrix}$$

正溴丁烷理论产量：0.2×137 = 27.4（g）

$$产率 = \frac{实际产量}{理论产量} \times 100\% = \frac{18\ g}{27.4\ g} \times 100\% = 66\%$$

8. 实验讨论

（1）在回流步骤中，上、中层液体为橙黄色，可能是混有少量溴所致，溴是由硫酸氧化溴化氢产生。

（2）粗产物中含有未反应的正丁醇和正丁基醚等副产物。这些杂质可以用浓硫酸洗涤除去。因为醇和醚能与浓硫酸反应，形成盐并溶解于浓硫酸中。

Chapter 1 Introduction of Organic Chemistry Experiments

Chemistry is an experimental science. During this organic chemistry experiment course, you will acquire a basic understanding of chemical reactions on the molecular level, and understand why certain conditions are required to get molecules to react with one another. At the same time, you will have the opportunity to learn the practical operation procedures of organic chemistry experiments in this experimental course, discover the scientific inquiry methods commonly used by organic chemists, think about the relationship between the theoretical knowledge of organic chemistry and the experimental phenomena, learn the skills in how to acquire and reasonably explain your experimental results, and enable you to understand, master and apply the knowledge points learned in organic chemistry class.

In general, the experimental skills you develop will help you in future experimental courses in chemistry and in other fields of science. Careful observation and documentation of data, teamwork and collaboration with others, and locating information about chemicals and chemical reactions in the literature are important skills that you will use in many aspects of life. Your laboratory work will give you the opportunity to exercise your critical thinking abilities, and get into the habit of rigorous scientific attitude and work style.

1.1 General Rules for Organic Chemistry Experiments

Before you begin your adventure in the organic chemistry experiments, it is important that you are aware of the important safety rules for your well-being as well as that of other students. It is essential that you follow the safety guidelines listed in this section and the safety notes included in each experiment. We urge you to read them carefully before you begin laboratory work.

1) Before the experiments

(1) Carefully read the experimental content to be performed before you come to the laboratory and complete a pre-lab report in your laboratory notebook.

（2）Carefully check the experimental instruments and drugs. If there is any shortage or damage， you should report to the experimental instructors immediately.

（3）Must wear lab-gowns，safety glasses or goggles and latex gloves at all times in the lab.

2）During the experiments

（1）Carefully listen to the instructors explaining the purpose，steps，performance of instruments and drugs， operating methods and matters needing attention.

（2）No hubbub， smoking or eating in the lab.

（3）Strictly carry out the operating procedures， carefully observe the experimental phenomena， make a good record of the experiments， analyze the experimental results according to the experimental process， and write the experimental reports.

（4）Pay attention to safety， do be careful and cautious when using corrosive， flammable， explosive and poisonous drugs. Do not panic， if there is an accident in the experiment， you should immediately report to instructors to deal with it.

（5）Protect the instruments and experimental materials， save water and experimental materials. Minimize the amount of chemicals you use and dispose of waste chemicals properly. Laboratory equipment and materials are not allowed to be taken out of the lab.

3）After the experiments

（1）After the experiments， under the guidance of the experimental instructors and the experimental management teachers， the laboratory equipment should be checked， and returned to the original places；dispose of the waste properly，clean the instruments properly， keep indoor objects tidy， and leave the laboratory with the permission of the teacher.

（2）Before leaving the chemical laboratory， check whether water， electricity， doors and windows are closed. Carefully and safely hold the inflammable， corrosive drugs and toxic substances.

1.2 General Knowledge for Organic Chemistry Laboratory Safety

The organic chemistry laboratory is a place where accidents can and do occur and where safety is everyone’s business. While working in the laboratory，you are protected by the instructions in an experiment and by the laboratory itself， which is designed to safeguard you from most routine hazards. However，neither the experimental directions nor the laboratory facilities can protect you from the worst hazard—your own or your neighbors’ carelessness.

Some laboratory accidents are general types, involving fires, explosions, electrical appliances, cuts, chemical burns, absorption through the skin or ingestion of toxic materials. The following precautions and treatments should be followed to minimize the chances of accidents.

1.2.1 The prevention and treatment of experimental hazards in organic chemistry laboratories

1. Fire

Fire hazards in organic chemistry laboratories are often considerable due to the quantities of volatile and flammable chemicals, particular solvents, which are commonly used. The vapor may drift to a distant ignition source and burn back to ignite the bulk of the liquid. An important rule is never to allow any vapor of a volatile chemical to escape into the open laboratory (in addition to fire hazards many vapors are toxic). Strict regulations should be applied to the total quantity of solvents which may be stored in a laboratory, and with the exception of small bench reagent bottles, storage must be in approved cabinets. If the spillage of solvents or the accidental release of flammable vapor occurs, the whole laboratory should be ventilated as soon as possible. Always check the gas valve and the gas light and keep them in good condition from leakage.

Keep calm and treat in time in the fire accidents; the following measures are generally taken.

(1) To prevent the fire from spreading. Immediately put out the fire in the vicinity, cut off the power supply, and remove the combustible away from fire.

(2) Put out the fire immediately according to the intensity of a fire. If the igniting area is small, use asbestos cloth or sand to cover the fire; if the igniting area is big, use the fire extinguishers. To extinguish fire caused by organic matter, do not use water; to an electrical fire, cut off the power at first, and then use dry ice fire extinguishers to put out the fire at once.

In general, when a fire breaks out, different methods should be adopted to extinguish it according to the cause of the fire and the circumstance around the fire site. No matter what kind of fire extinguishing equipment is used, the fire should be put out from around the fire source to the center, and the nozzle of the fire extinguisher should be aimed at the bottom of the flame. Don't hesitate in the rescue process. Types and characteristics of common fire extinguishers are listed in Table1.1.

Table 1.1 Types and characteristics of common fire extinguishers

English names	Shorthand	Characteristics
foam fire extinguisher	MP	suitable for use in extinguishing oil fire，unsuitable for use in extinguishing electrical fire
carbon dioxide fire extinguisher	MT	suitable for use in extinguishing electrical fire，fire caused by oil and water-repellent chemicals in a small area
dry powder fire extinguisher	MF	suitable for use in extinguishing fire caused by the electrical equipment，oil，combustible gas，precision instruments，books and materials，water-repellent chemical fire

2. Explosion

（1）Never heat a closed system! Never completely close off an apparatus in which a gas is being evolved. Always provide a vent in order to prevent an explosion.

（2）Do not make flammable and explosive gases close to the source of fire.

（3）When using ethers such as sulfuric ether，it is necessary to check the existence of peroxides. If the presence of peroxides is found，ferrous sulfate should be used immediately to remove them，and ethers should be used in a well-ventilated place or in a fume hood.

（4）For explosive solids，such as heavy metal alkynes，metal salts of picric acid and trinitrotoluene，no heavy pressure or impact is allowed to cause explosion. These residues are dangerous and must be carefully destroyed.

3. Electrical safety

Many accidents in laboratories are caused by the malfunctioning of electric appliances and by thoughtless handling. Therefore，do not touch the plug with wet hands or with wet objects，electric equipment should be carefully inspected to check that the plug has been correctly fitted to ensure that it is in good condition with no loose wires or connections，and is properly earthed. After the experiments，the power supply should be turned off and the power plugs should be unplugged.

4. Burns and cuts

Learn the location of the first aid kit and the materials it contains for the treatment of simple cuts and burns.

When heating glassware，do not touch the hot spot. Do not put hot glassware on a bench where someone else might pick it up. Steam and boiling water cause severe burns. Handle containers of boiling water very carefully. Apply cold water for 10-15 min to minor heat burn. Seek immediate medical attention for any extensive burn.

Cut injuries are the common accidents. After an injury, we should carefully observe whether there are glass fragments in the wound. If so, we should first remove the glass fragments from the wound. Minor cuts may be treated by ordinary first-aid procedures. As to severe bleeding, attempt to stop the bleeding with compresses and pressure, and arrange for emergency room or infirmary treatment at once.

5. Chemical burns

The first thing to do if any chemical is spilled on your skin, unless you have been specifically told otherwise, is to wash the area well with water for 10-15 min. This treatment will rinse away the excess chemical reagent. For acids, bases, and toxic chemicals, thorough washing with water will save pain later. Skin contact with a strong base usually does not produce immediate pain or irritation, but serious tissue damage (especially to the eyes) can occur if the affected area is not immediately washed with copious amounts of water. If the chemical burns are minor, apply burn ointment; seek immediate medical treatment for any serious chemical burns.

If a chemical gets into your eyes, immediately go to the eye wash station and wash your eyes with a copious amount of slightly warm water. Hold your eyes open to allow the water to flush the eyeballs for 10-15 min. Seek medical treatment immediately after using the eye wash for any chemical splash in the eyes.

6. Poisoning

(1) Don't smell the reagent directly by nose.

(2) Don't eat, drink or taste any reagent in the lab.

(3) Toxic residues must be handled properly and effectively. No littering!

(4) You must put on rubber gloves when handling with toxic reagents and treating with them in a ventilation cabinet. Keep them away from your mouth or skin, and never pour them into the drain. Wash your hands immediately after an operation and move to an area where you can breathe fresh air and rest.

(5) When you operate an experiment in the ventilation cabinet, don't stretch your head into the cabinet. It's important to keep your head out of the front panel.

（6）If the symptoms of poisoning appear，you should receive medic mediate treatment. If the toxic chemicals have splashed into the mouth，spit out immediately，and rinse mouth with plenty of water. If they have been already swallowed，the first aid should be done according to the nature of the poison as follows：①Acid. Drink plenty of water first，and then take aluminum hydroxide，egg white or milk. Don't take vomiting agents.②Base. Drink lots of water first，and then take vinegar，sour juice，egg white or milk. Don't take vomiting agents.③Irritating or nerve poisons. The first aid is to promote a vomiting as quickly as possible. Drink milk or egg white to dilute the poisons at first，and then drink a spoonful of magnesium sulfate（about 30 g）dissolved in a glass of water. Sometimes you can promote vomiting by pressing the throat with a finger. Seek immediate medical attention.④Inhalation of gaseous poisons. Move to a safe area quickly，unlocking the collar and buttons. If inhale chlorine or bromine gas，gargle with dilute solution of $NaHCO_3$.

1.2.2 Disposal of lab waste

The proper disposal of inorganic and organic waste chemicals is one of the biggest responsibilities that you have in the organic laboratory. Your actions，and those of your lab mates，can minimize the environmental impact and even reduce financial cost to your school for handling the waste chemicals that are necessarily produced in the experiments you do. Thus，each student must assume the responsibility for seeing that her or his spent chemicals go into the appropriate container. It is essential that everyone cooperate in disposing of materials correctly in order to protect our environment.

（1）Waste solvents should be placed in suitable containers and appropriately labelled，but indiscriminate mixing of solvents must be avoided. Halogenated solvents in particular should be kept apart from other solvents.

（2）A few materials used in the lab can be put into an ordinary wastebasket or other similar solid waste receptacles，such as innocuous waste solids，common paper or cloth.

（3）The broken glassware instruments or droppers should be placed in the designative bin.

（4）Toxic solids should be sealed in a plastic bag and placed in a separate bin.

（5）All bins should be clearly labelled.

（6）All waste containers should be kept closed when not in use.

1.3 Common Glass Instruments and Apparatuses for Organic Chemistry Experiments

1.3.1 Common glassware

As shown in Figure 1.1, some typical sets of lab glassware for organic chemistry experiments are listed.

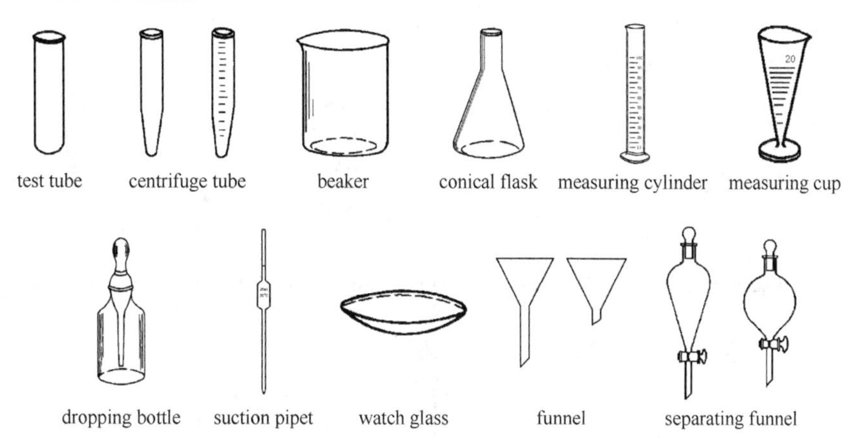

Figure 1.1 Common glassware for organic chemistry experiment

1.3.2 The standard ground glass instrument

In organic chemistry experiments, standard ground glasses are usually used. Because of the different uses of glass instruments, the sizes of standard ground mouth are also different. Standard taper joints come in a number of sizes and are designated by the symbol followed by two sets of numbers separated by a slash, as in 10/30. The first number is the diameter of the joint in millimeters at its widest point, and the second is the length of the joint in millimeters. A standard taper joint that is designated as 10/30 therefore has a widest diameter of 10 mm and a length of 30 mm. Standard ground glass joints are used in laboratories to quickly and easily fit leak-tight apparatus together from commonly available parts.

Common standard ground glass instruments are shown in Figure 1.2.

1.3.3 Common apparatus

In an organic chemistry experiment, whether the experimental device is assembled correctly or not is closely related to the success of the experiment.

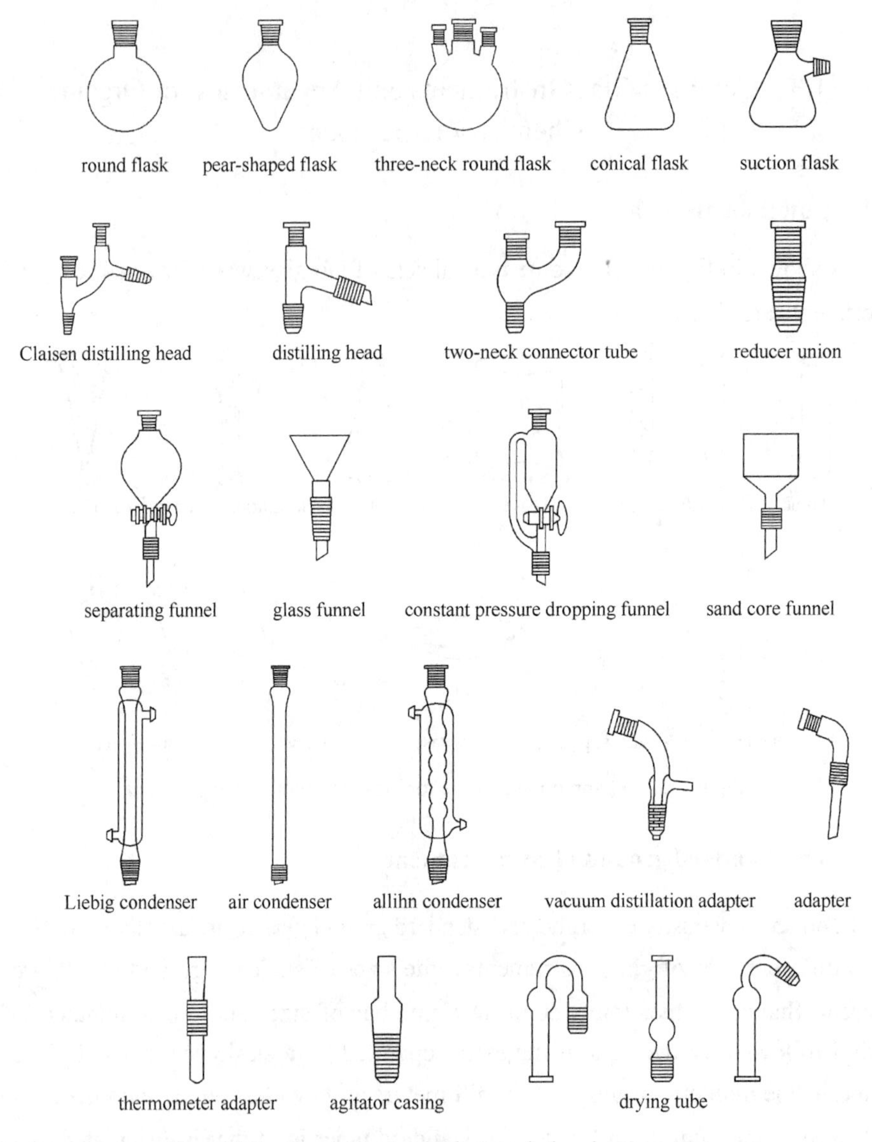

Figure 1.2 Common standard ground glass instruments

Firstly，when assembling a set of devices，the glass instruments and accessories selected should be clean，otherwise the yield and quality of the products will be affected. Secondly，the equipment chosen should be appropriate. For example，in experiments requiring heating，if a round-bottom flask is selected，the volume of the flask should be about 1/2 of the total volume of the reactant and not more than 2/3 at most. Finally，the reactor position should be selected according to the heat source，and the reactor should be fixed with iron clamps. Then the reactor should be assembled one by one in the order of the "bottom-to-top" and "left-to-right" principle. When disassembling, the reactor should be disassembled one by one in the opposite order.

Assemble the apparatus following: tight, correct, regular and stable. Correct experimental device should not only be reasonable and practical, but also neat and beautiful.

Common apparatuses are shown in Figure 1.3, Figure 1.4 and Figure 1.5.

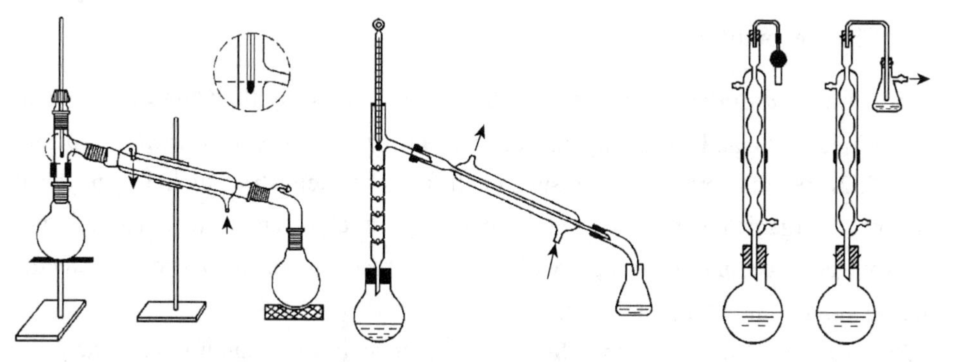

Figure 1.3 Simple distillation apparatus

Figure 1.4 Fractional distillation apparatus

Figure 1.5 Reflux apparatus

1.4 Organic Chemistry Literature

1.4.1 Laboratory handbooks

(1) *CRC Handbook of Chemistry and Physics*: Annual edition, CRC Press, Boca Raton, FL. The CRC is probably the best-known single-volume handbook, and it contains a wealth of useful data in all areas of chemistry and physics. The most useful part for organic chemists is the section headed "Physical Constants of Organic Compounds", which provides formulas, structures, molar masses (molecular weights), densities, refractive indexes, solubility, colors, and melting and boiling points for about 11000 compounds.

(2) *The Merck Index of Chemicals and Drugs* (14th ed). It is published by Merck Company, Rahway, NJ. It gives a concise summary of the physical, chemical and pharmacological properties of more than 10000 compounds, including pharmaceuticals, organic chemicals and reagents, inorganic substances, agricultural chemicals, and naturally occurring substances.

(3) *Beilstein Handbook of Organic Chemistry*. It is the largest compilation of information on organic compounds. Earlier volumes were published in Germany, but

since 1984 it has been published in English. It is a reference work containing information and data relating to the structure，preparation and properties of organic compounds reported in the primary literature. It contains records on almost 10 million organic substances.

1.4.2 Journals

1. Chinese journals

Acta Chimica Sinica（monthly）published in Chinese by the Chinese Chemical Society since 1933 and originally named *Journal of the Chinese Chemical Society* before 1952，reports new research results in all fields of chemistry，including physical chemistry，inorganic chemistry，organic chemistry，analytic chemistry，and polymer chemistry，etc. Accounts are summary of significant recent work and new developments from research groups of principal authors.

Progress in Chemistry（monthly）is a peer-reviewed monthly journal. It is sponsored by the Chinese Academy of Sciences（CAS）and the National Natural Science Foundation of China（NSFC）. It provides a forum to publish review papers of specialized topics covering the full spectrum of chemistry in Chinese or English，with emphasis on those topics of the emerging research areas. The reviews provide comprehensive information，including recent advances，development trends，as well as critical assessments about the subjects. The readers interested are：researchers and students in chemistry and related areas，and policy decision-makers. Most members of the editorial board are famous scientists.

2. English journals

Journal of the American Chemical Society（J. Am. Chem. Soc.，JACS），weekly，founded in 1879，is the flagship journal of the American Chemical Society and one of the world's preeminent journals in all of chemistry. This periodical is devoted to the publication of fundamental research papers and publishes approximately 19000 pages of articles，communications and perspectives a year. JACS provides research essential to the field of chemistry.

Angewandte Chemie International Edition（Angew. Chem. Int. Ed.），weekly，and its German version Angewandte Chemie are owned by the Gesellschaft Deutscher Chemiker（German Chemical Society）and are published by Wiley-VCH. It is a leading journal for all fields of chemistry. Both editions of the journal have 52 issues in print and online (in the Wiley Online Library) per year, all articles are available online weeks before they appear in an issue (online and print) .

Chemical Reviews (Chem. Rev.), biweekly, is one of the most highly regarded and highest-ranked journals covering the general topic of chemistry. The mission of Chemical Reviews is to provide comprehensive, authoritative, critical and readable reviews of important recent research in organic, inorganic, physical, analytical, theoretical and biological chemistry. In addition to the general reviews, the journal has published since 1985 periodic thematic issues focusing on a single theme or direction of emerging researches.

1.4.3 Chemical abstracts

Chemical Abstracts (CA), which was first published by the American Chemical Society in 1907, is the world's largest index. Annually abstracts over a million documents are drawn from about 16000 technical journals as well as from patents, reviews, technical reports, monographs, conference proceedings, symposia, dissertations and books. *Chemical Abstracts* condenses the content of journal articles into abstracts and indexes the abstracts by research topics, author's names, chemical substances or structures, molecular formulas, and patent numbers. Each chemical compound is assigned a number, called a registry number, which can facilitate the finding references to the compounds.

Now, CA is now available electronically in several different ways. Chemical Abstract Services (CAS), the publishers of Chemical Abstracts, provides a number of databases. The newest of these databases, called SciFinder Scholar, is an excellent search engine.

1.4.4 Internet resources

There are an increasing number of online databases readily available to the organic chemist. It is likely that all researchers will require skills to conduct searches of such databases in future. Therefore, the course in organic chemistry should provide an introduction to online searching techniques in order to provide the basis for development later in the students' career.

(1) Online libraries.

National Digital Library of China http://www.nlc.cn/;

Jilin University Library http://lib.jlu.edu.cn/portal/index.aspx/.

（2）Online periodical resources.

China National Knowledge Infrastructure（CNKI）　http://www.cnki.net/;

Acta Chimica Sinica　http://sioc-journal.cn/Jwk_hxxb/CN/volumn/current.shtml;

Progress in Chemistry　http://manu56.magtech.com.cn/progchem/CN/volumn/home.shtml.

（3）Online patent resources.

China and global Patent Examination Information Inquiry

http://cpquery.sipo.gov.cn/.

（4）Database resources.

Chem Blink　http://www.chemblink.com/indexC.htm;

Chemistry Database　http://www.organchem.csdb.cn/.

1.5 Experimental Preview Records and Laboratory Reports on Organic Chemistry

1.5.1 Preview report

This is an important part of any chemical experiment, every student should prepare a lab notebook and must carefully and fully preview each experiment before start, since careful observation, allied to accurate reporting, is the very essence of any scientific exercise.

The specific requirements of the preview are as follows.

（1）Preview what are requested by the instructions for each experiment in this book.

（2）Understand the reaction principle（main reaction and side reaction）.

（3）Write the physical constants of the main reagents and products, and specifications and quantities（g, mL, mol）of the main reagent; and list all glassware for need.

（4）List general procedures of the experiment and any chemical hazards it might present.

（5）List a table for recording the data.

1.5.2 Lab records

The guiding principle in writing up an experiment is to record all the details which would enable another person to understand what was done and to repeat the entire experiment exactly without prior knowledge. Thus, during an experiment, the student should stick to standard operations, observe carefully and think actively, and also should record timely the experimental phenomena (such as the heating conditions, the changes of the color or pH value, precipitation and gas emitting, etc.) and data (such as room temperature, reaction temperature, etc.). It is important that numerical results such as yields, melting points and boiling points, etc., are recorded directly into the notebook. The writing up of all laboratory work must be done at the time of the work. Writing notes and recording results by later memories are never allowed. The lab record should be real, concise and clear. After the experiment is finished, the original records and the products obtained should be presented and checked by teacher before leaving.

In organic synthesis experiments, the yield and quality of the products are usually one of the important indicators for evaluating an experimental result and evaluating students' experimental skills. The theoretical mass in an organic reaction is the mass of the product which would be obtained if the reaction had proceeded to completion according to the chemical equation. The mass of product is isolated from the reaction. The percentage yield may be expressed thus:

$$\text{Yield} = \frac{\text{actual mass of the product}}{\text{theoretical mass of the product}} \times 100\%$$

1.5.3 Experiment report

Experiment report is an important part of the experimental practice for a student. A complete final report should include the experiment objectives (what to do), the experiment principles (why to do), the experiment procedures with the reagents and apparatus you have used (how to do), and the experiment results you have obtained. The final summary and conclusion on your experimental results are especially important and the discussions on improvement advice or solution to some experimental failures are also required. A report sample is given as follows.

Experiment Title

(1) Objectives of the experiment (main techniques introduced for the experiment).

(2) Experimental principles (chemical reaction, chemical reaction equation or reaction mechanism).

(3) Physical constants of the main reagents and the product.

(4) The specification and dosage of the main reagents and the experimental apparatus.

（5）Experimental apparatus.

（6）Experimental procedures and observations（color changes，solubility，precipitation etc.）.

（7）Experimental conclusion（yield，m.p，b.p etc.）.

（8）Experimental discussion.

Sample：

Preparation of *n*-bromobutane

1.【Objectives】

（1）To learn the principles and methods of preparing *n*-butyl bromide from *n*-butyl alcohol.

（2）To master the techniques of reflux with a gas trap apparatus，simple distillation apparatus and extraction.

2.【Principles】

Main reactions：

$$NaBr + H_2SO_4 \longrightarrow HBr + NaHSO_4$$

$$CH_3CH_2CH_2CH_2OH + HBr \xrightarrow{H_2SO_4} CH_3CH_2CH_2CH_2Br + H_2O$$

3.【Physical constants of the main reagents and products】

Name	Relative molecular mass	Character	Relative density	Melting point/℃	Boiling point/℃	Solubility/（g/100 mL）		
						Water	Ethanol	Aether
n-butyl alcohol	74.12	colorless and transparent liquid	0.8098	88.9	117.25	7.9*	∞	∞
n-bromobutane	137.03	colorless and transparent liquid	1.2758	112.4	101.6	insoluble	∞	∞

* 20℃ room temperature

4.【Materials】

n-butyl alcohol C.P. 18.5 mL（0.20 mol），sodium bromide C.P. 25 g（0.24 mol），concentrated sulfuric acid L.R. 29 mL（0.54 mol），saturated sodium bicarbonate solution，anhydrous calcium chloride C.P. 2 g（0.54 mol）.

5. 【Apparatus】

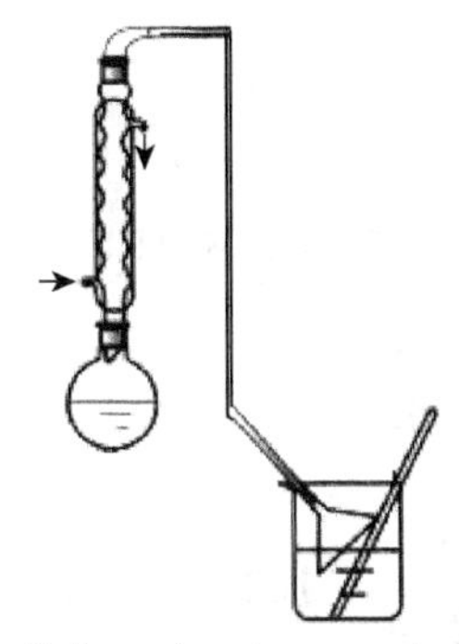

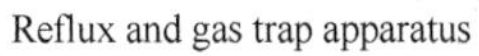

Reflux and gas trap apparatus

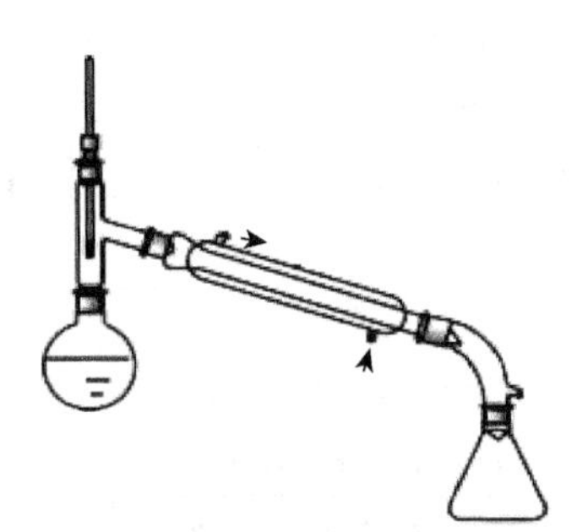

Simple distillation apparatus

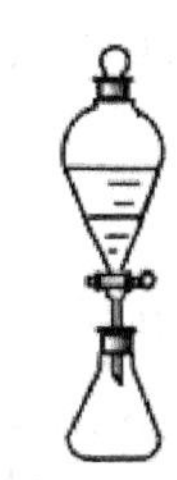

Extraction apparatus

6. 【Procedure】

	Procedure	Experimental phenomena
Reflux	Place 20 mL of water into a 150 mL round-bottom flask and slowly add 29 mL of concentrated sulfuric acid with continuous swirling. Cool the mixture to room temperature and add 18.5 mL of *n*-butyl alcohol and 25 g of sodium bromide. Add a stir bar to the mixture and place the round-bottom flask in a heating mantle, clamp it securely, and connect it with a reflux apparatus. Be sure not to merge the inverted funnel of the gas trap into aqueous solution in order to avoid inverse suction. Heat the mixture in the heating mantle, swirling frequently. Continue heating the mixture for 60 min. Then stop heating and allow the apparatus to cool before disconnecting the reflux apparatus	Exothermic. Liquid stratification in reaction. The upper layer that soon separates is the *n*-bromobutane, Light yellow to orange
Simple distillation	Remove the condenser with the gas trap and assemble a simple distillation apparatus. Mount a distillation head in the round-bottom flask, and set the Liebig condenser for downward distillation through a bent or vacuum adapter into a 50 mL conical flask. Distill *n*-bromobutane until the distillate appears to be clear	At first, the distillate was milky-white oily substance, and then the oily substance decreased
Extractions	Pour the distillate in a 50 mL separating funnel and add the equivalent volume of water to wash the mixture. Stopper the funnel and shake it, venting occasionally. Allow the layers to separate. The crude *n*-bromobutane layer should be lower. Drain the lower layer through the stopcock into another dry separating funnel. Add the equivalent volume of concentrated sulfuric acid to the second funnel and shake the mixture. Allow the layers to separate. Drain and discard the lower sulfuric acid layer. Extract the organic layer with water again, and then with saturated aqueous sodium bicarbonate. Note that the upper water phase should be separated from the lower organic phase as completely as possible in the last extraction	Layered. Compared with water, organic phase is in the lower layer; compared with sulfuric acid, organic phase is in the upper layer. The crude product is brown yellow liquid
Drying	Drain the organic layer into a dry Erlenmeyer flask. Add about 2 g of anhydrous calcium chloride to dry the solution for about 20 min until the liquid is clear	Turbid to clear
Simple distillation	Pour the dried liquid into a dry round-bottom flask carefully, remaining the drying agents in the Erlenmeyer flask. Add a stir bar and distill the crude *n*-bromobutane. Collect the distillate boiling in the range of 99-103℃	The temperature is quickly raised to 99℃, stabilized at 101-102℃, and finally raised to 103℃
Yield determination	Weigh the product and calculate the yield	Colorless and transparent liquid

7.【Yield】

Experimental weight of product：18 g

Theoretical weight of product：

$$CH_3CH_2CH_2CH_2OH + HBr \xrightarrow{H_2SO_4} CH_3CH_2CH_2CH_2Br + H_2O$$

1　　　　　　　　1

0.2　　　　　　　0.2

Theoretical weight：0.2×137 = 27.4（g）

$$\text{Yield} = \frac{\text{actual mass of the product}}{\text{theoretical mass of the product}} \times 100\% = \frac{18\text{g}}{27.4\text{g}} \times 100\% = 66\%$$

8.【Discussion】

（1）In the reflux procedure，the upper and middle liquids are orange yellow，which may be caused by the mixture of small amounts of bromine，which is produced by the oxidation of hydrogen bromide with sulfuric acid.

（2）The crude products contain unreacted *n*-butanol and by-products such as *n*-butyl ether. These impurities can be removed by washing with concentrated sulfuric acid because alcohols and ethers can react with concentrated sulfuric acid to form salts and dissolve in it.

第2章　有机化学实验基本操作

2.1　玻璃仪器的洗涤与干燥

2.1.1　玻璃仪器的洗涤

化学实验中经常使用各种玻璃仪器，如果使用不洁净的仪器，往往由于污物或杂质的存在而得不到正确的结果，因此玻璃仪器的洗涤是化学实验中一项重要的操作技能。

玻璃仪器的洗涤方法很多，应根据实验要求、污物的性质和沾污的程度来选择合适的洗涤方法。

玻璃仪器上的大多数化学残留物可以用毛刷、实验室专用去污粉和水清洗除去。

由浓硫酸和铬酐或重铬酸钾制成的铬酸洗液是一种有效的清洗液，但由于它含有强氧化性酸，所以必须小心使用。玻璃仪器清洗干净后，将铬酸溶液倒入指定的洗液回收瓶中，不要倒入水槽中，以免腐蚀下水道。

另一类有效清洗液是有机溶剂，如丙酮，可与水混溶，它能溶解大部分有机残留物，因此常用于清洁玻璃仪器。

2.1.2　玻璃仪器的干燥

（1）晾干。对于不急用的仪器，可将仪器倒插在实验室的干燥架上晾干。

（2）吹干。将仪器倒置控去水分，并擦干外壁，用电吹风的热风将仪器内残留水分赶出。

（3）烘箱干燥。湿的玻璃器皿可以在120℃的烘箱中加热20 min来干燥。用钳子把干燥的玻璃器皿从烘箱中取出，让它冷却到室温，然后使用。

（4）有机溶剂干燥。在用水洗净的仪器内加入少量有机溶剂（如丙酮、无水乙醇等），转动仪器，使仪器内的水分与有机溶剂混合，倒出混合液（回收），仪器即迅速干燥。

必须注意，在化学实验中，某些情况下并不需要将仪器干燥，如量器、容器等，使用前先用少量溶液冲洗两三次，洗去残留水滴即可。带有刻度的计量容器不能用加热法干燥，否则会影响仪器的精度，如需要干燥时，可采用晾干或冷风吹干的方法。

2.2　加热与冷却

2.2.1　加热

在室温下，一些有机反应很难进行或反应速度很慢。为了提高反应速率，通常需要加热条件，如蒸馏和升华过程需要加热。有机化学实验室常用的热源有煤气灯、电热套、电热板等，一般不直接加热，为了避免直接加热可能出现的安全问题，通常根据具体情况采用以下间接加热方法。

1. 水浴

对于沸点低于 100℃的易燃液体溶液，必须使用配备有恒定液位装置的电热水浴锅加热，如图 2.1 所示，有时为了方便，常用规格较大的烧杯等替代。加热温度低于 90℃时，容器可直接浸入水浴中。采用沸水浴或蒸汽浴可达到 95℃左右的温度。在蒸汽浴的操作中，容器悬浮在水上，水被蒸汽加热。由于水的蒸发，在操作过程中应补加入一定量的热水，使水浴液面保持高于反应容器内溶液的上表面。

图 2.1　水浴装置

2. 油浴

油浴适用于 100～250℃加热，温度可以很容易地控制在一定范围内，容器内的反应物受热均匀。反应体系温度一般比油浴温度低 20℃左右。与水浴一样，油浴上液面应保持高于反应物溶液的上表面。

使用油浴时，操作人员必须特别注意防火。浴油发烟严重时应立即停止加热。一旦着火，操作人员应首先关闭电源或煤气灯，然后清除周围的可燃物，最后用石棉覆盖油浴。为避免过热，应始终在浴缸中放置温度计。不允许将水引入油浴缸中，因热水可能会飞溅出来。烧瓶应在油浴缸上方悬放几分钟，然后从油浴中取出后用抹布擦拭。

油浴所能达到的最高温度取决于所使用浴油的种类。有机化学实验室常用的几种导热油如下：

（1）甘油。可以加热到 140～150℃，温度过高时会分解，蒸气的气味令人不快。由于甘油具有很高的亲水性，在长时间放置后，使用前应加热甘油以除去吸收的水分。

（2）植物油。如菜油、蓖麻油和花生油，可以加热到 220℃。常加入 1%对苯二酚等抗氧化剂以提高油的热稳定性，便于久用。温度过高时会分解，达到闪点可能燃烧，所以使用时要十分小心。

（3）石蜡。能加热到 200℃左右，保存方便。冷却到室温则成为固体，因此使用后应立即从槽中取出石蜡。

（4）液状石蜡。可加热到 200℃左右，它具有较高的热稳定性，但在高温下更容易燃烧。

（5）硅油。硅油在 250℃时仍较稳定，不会明显分解和变色，透明度好，但价格稍贵。

3. 砂浴

当所需的加热温度远远高于上述所列温度时，通常可以使用砂浴。其最高工作温度可达 350℃。干净干燥的砂子被放置在砂浴铁盒中，容器被半埋在砂浴铁盒中。砂浴的缺点是导热性差，但散热快，导致温度分布不均匀。因此，砂浴铁盒底部的沙子应该稍微薄一些，但容器周围的砂子要稍微厚一些。应始终将温度计插入砂浴盒中，以检测温度，其中水银球应靠近容器。

4. 空气浴

空气浴法是一种非常价廉、方便的小型蒸馏烧瓶加热方法，对沸点在 80℃以上的液体原则上均可采用空气浴加热。当使用煤气灯作为热源时，必须使用远离烧瓶约 1 cm 的耐热石棉网隔开。最简单的空气浴如图 2.2 所示。然而，这种方法容易由于周围空气流动引起加热的水平波动。它不适用于低沸点液体的回流或减压蒸馏。

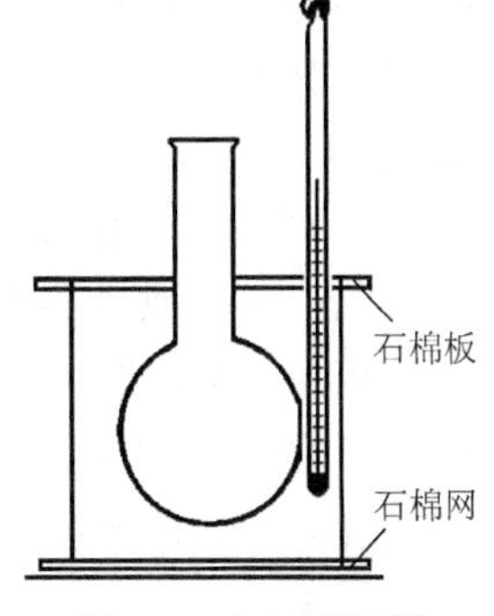

图 2.2　空气浴装置

5. 电热套

电热套或称恒温套，是一种实验室常用加热设备。它由一个加热元件组成，该加热元件被密封在针织玻璃纤维内，使之更加安全。与其他加热装置（如加热板或煤气灯）不同，玻璃器皿容器可直接与电热套接触，而不会增加玻璃器皿破碎的风险，因为电热套的加热元件与反应容器绝缘，以防止温度梯度过高。由于电热套不是明火加热，因此可以加热和蒸馏易燃有机物，也可加热沸点较高的化合物。电热套加热温度可从室温到 200℃左右，适用加热温度范围较广。

2.2.2　冷却

放热反应进行时，常伴随大量的热产生，使反应温度迅速升高，如果控制不当，往往会引起反应物的蒸发，使其逸出反应器，也可能引发副反应，甚至引起爆炸。为了将温度控制在一定范围内，需要适当冷却。因此，在有机化学实验中，经常需要采用一定的冷却剂进行冷却操作，在一定的低温条件下进行反应、分离、提纯等。冷却剂是根据冷却需要达到的温度和带走的热量来选择的。

（1）水。所有冷却剂中，水成本最低，且热容量高，故为常用的冷却剂。但随着季节的不同，其冷却效率变化较大。此外，在回流、蒸馏等操作中也常用水作冷却剂，以冷却高温的气体或液体。

（2）冰-水混合物。某些反应需在低于室温的条件下进行，则可用水和碎冰的混合物作冷却剂，冰-水浴可冷却至 0～5℃，由于能和容器更好地接触，其冷却效果要比单用冰块好。如果水的存在并不妨碍反应的进行，则可以把碎冰直接投入反应物中，这样能更有效地保持低温。

（3）冰-盐混合物。如果需要将反应混合物保持在 0℃以下，常用碎冰和无机盐的混合物作冷却剂。制作冰-盐浴时，应把盐研细，然后和碎冰按一定比例均匀混合。在实验室中，最常用的冷却剂是碎冰和食盐的混合物，一般能将反应物冷却至–21～–5℃。

（4）干冰（固体二氧化碳）。干冰可冷却至–60℃以下。如果将干冰加到乙醇、丙酮等溶剂中，可冷却至–78℃，但加入时会猛烈起泡。使用该类冷却剂时，应将冷却剂置于杜瓦瓶（广口保温瓶）中或其他绝热效果好的容器中，以保持其冷却效果。

（5）液氮。液氮可冷却至–196℃。为了保持冷却效果，应将这种冷却剂置于杜瓦瓶中或其他绝热效果好的容器中。

应当注意，当冷却温度低于–38℃时，不能使用水银温度计，因为水银的凝固点为–38.9℃，测定温度时应采用内部添加少许颜料的有机液体（如乙醇、甲苯、正戊烷等）低温温度计。

2.3　重结晶与过滤

2.3.1　重结晶

重结晶是提纯固体化合物常用的方法之一。固体化合物在溶剂中的溶解度随温度变化而改变，一般温度升高，溶解度增加，反之则溶解度降低。如果把固体化合物溶解在热的溶剂中制成饱和溶液，然后冷却至室温或室温以下，则溶解度下降，原溶液变成过饱和溶液，这时就会有结晶固体析出。利用溶剂对被提纯物质和杂质的溶解度的不同，使杂质在热过滤时被滤除或冷却后留在母液中与结晶分离，从而达到提纯的目的。

重结晶的操作可分为六个步骤：溶剂选择和溶质溶解（即在沸点处制备接近饱和的溶液）、溶液脱色、热过滤、结晶析出、晶体收集和洗涤、产品干燥。

1. 溶剂选择和溶解溶质

在进行重结晶时，选择理想的溶剂是一个关键，理想的溶剂必须具备下列条件：

（1）不与被提纯物质发生化学反应。

（2）在较高温度时能溶解多量的被提纯物质，而在低温时只能溶解很少量的该种物质。

（3）对杂质溶解非常小或者非常大（前一种情况是杂质在热过滤的时候被滤去；后一种情况是杂质留在母液中不随被提纯物晶体一同析出）。

（4）晶体产出量较高，晶形好。

（5）无毒或低毒，便于操作。

（6）价廉易得。

如果两种或两种以上的溶剂都可以适用于重结晶，则溶剂的最终选择将取决于易操作性、毒性、易燃性和成本等因素。具体重结晶的溶剂选择可参考化学文献、手册，或根据“相似相溶”原则，通过预实验选定一种或多种混合溶剂。

溶剂选择步骤如下：将少量不纯溶液放在几个小试管中（通常每管约 0.2 g），然后分别向试管中添加 0.5 mL 不同的溶剂。将溶液加热至全部溶质溶解，然后冷却至室温。得到晶体量最多的溶剂通常被认为是最理想的溶剂。如果溶质仍未完全溶解在 3 mL 热溶剂中，则该溶剂不适用。如果溶质在热溶剂中溶解，但在溶液冷却时没有结晶，即使用玻璃棒轻刮溶液表面下的试管内壁也无晶体析出，说明待纯化的溶质在该溶剂中具有较大的溶解性，该溶剂也不适用。

如果发现溶质在一种溶剂（称为较好溶剂）中过于可溶，而在另一种溶剂（称为较差溶剂）中过于不可溶，则可经常使用混合溶剂或“溶剂对”进行重结晶，效果良好。当然，这两种溶剂必须完全混溶。常用的溶剂对有：乙醇和水、乙醇和二乙醚、乙醇和丙酮、二乙醚和石油醚、苯和石油醚等。通常，该类溶剂对溶解在化学结构上与之类似的物质。

重结晶通常在锥形烧瓶或烧杯中进行。当使用具有高可燃性和挥发性的有机溶剂时，必须使用带有回流冷凝器的合适圆底烧瓶。将要重结晶的物质放入适合的烧瓶中，加入足够的溶剂（少于所需量）以浸没溶质和数个沸石，然后在水浴（如果溶剂在 90℃下沸腾）或电热套上加热烧瓶，直到溶剂沸腾。用玻璃棒搅拌混合物加快溶解。逐渐加入溶剂，使混合物保持沸腾，直到所有溶质溶解。

在重结晶中，若要得到较纯和较高收率的产品，必须特别注意溶剂的用量。溶剂的用量需从两方面考虑，既要防止溶剂过量造成溶质的损失，又要考虑到热过滤时因溶剂的挥发、温度下降使溶液变成过饱和，造成过滤时在滤纸上析出晶体，从而影响收率。因此溶剂用量不能太多，也不能太少，一般比需要量多 15%～20%。

2. 溶液脱色

有时，有机反应会生成颜色深的高相对分子质量副产物，这些溶液中的有色杂质可以被活性炭吸附。溶液脱色过程中如果活性炭加入太少，过滤后溶液仍会有色，则需要重复脱色操作；如果活性炭加入太多，除了杂质外，还会吸收一些产品。将少量（溶质量的 1%～5%）活性炭逐渐加入到显色溶液中，将溶液煮沸 5～10 min，如步骤 3 所述，通过过滤去除活性炭。

3. 热过滤

为了除去不溶性杂质必须趁热过滤。热过滤操作说明详见 2.3.2 部分。

4. 结晶析出

将上述热过滤后的溶液静置，自然冷却，结晶慢慢析出。结晶的大小与冷却的温度有关，一般迅速冷却并搅拌，往往得到细小的晶体，表面积大，表面吸附杂质较多。如将热滤液慢慢冷却，析出的结晶较大，但往往有母液和杂质包在结晶内部。因此，要得到纯度高、结晶好的产品，还需要摸索冷却的过程，但一般只要让热溶液静置冷却至室温即可，或放置在冰水浴中进一步冷却。有时遇到冷却后也无结晶析出时，可用玻璃棒在液面下摩擦器壁或投入该化合物的结晶作为晶种，促使晶体较快地析出。

5. 晶体收集和洗涤

析出的晶体常用减压过滤使其与母液分离。

6. 产品干燥

减压过滤后的结晶因表面还有少量溶剂，为保证产品的纯度，必须充分干燥。根据结晶的性质可采用不同的干燥方法，如自然晾干、红外灯烘干和真空恒温干燥等。

完全干燥后的结晶，称其质量，测熔点，计算产率。如果纯度不符合要求，可重复上述操作，直至熔点符合为止。

2.3.2　过滤

过滤是利用多孔介质（如滤纸、滤布等）进行分离溶液中悬浮固体颗粒的方法。通过过滤，固体物质留在滤纸上，液体通过过滤器进入接收容器，由过滤器获得的溶液称为滤液。通常，过滤方法有常压过滤、减压过滤、热过滤。

待过滤混合物的温度、黏度和过滤压力以及物料的状态都会影响过滤速度。溶液黏度越大，过滤器的速度越慢；减压过滤的速度比常压过滤快；当是胶状固体时，必须加热破坏它，以避免它通过滤纸。总之，在选择不同的过滤方法时，需要考虑各种因素。

1. 常压过滤

常压过滤（普通过滤）是在大气压力下用普通过滤漏斗过滤的方法。过滤的驱动力是物质自身重力，所以它是最简单、常用方法，但速度慢。这种方法通常用于去除不溶性物质。有时它可以用来分离晶体中较大的颗粒，吸收性差的晶体产品。该方法也用于分析化学实验。

根据沉淀的性质选择滤纸的种类：选择“慢速”滤纸过滤细晶沉淀，选择“快速”滤纸过滤胶体沉淀，选择“中速”滤纸过滤粗晶沉淀。依漏斗大小选择滤纸尺寸。

2. 减压过滤

减压过滤也称吸滤或抽滤，其装置如图 2.3 所示。水泵带走空气使抽滤瓶中压力低于大气压，布氏漏斗的液面与瓶内形成压力差，从而提高过滤速度。在水泵和抽滤瓶之间往往安装安全瓶，以防止因关闭水泵或水流量突然变小时自来水倒吸入抽滤瓶，如果滤液有用，则被污染。

如图 2.3 所示，陶瓷布氏漏斗（漏斗颈部的斜面应朝向过滤瓶的侧臂）装有橡胶塞，橡胶塞位于抽滤瓶颈部：抽滤瓶的侧臂使用厚壁橡胶管连接到安全瓶，然后连接到水泵；安全瓶上装有一个旋塞来调节装置中的压力，防止因水压突然下降可能导致水泵水倒吸污染滤液。

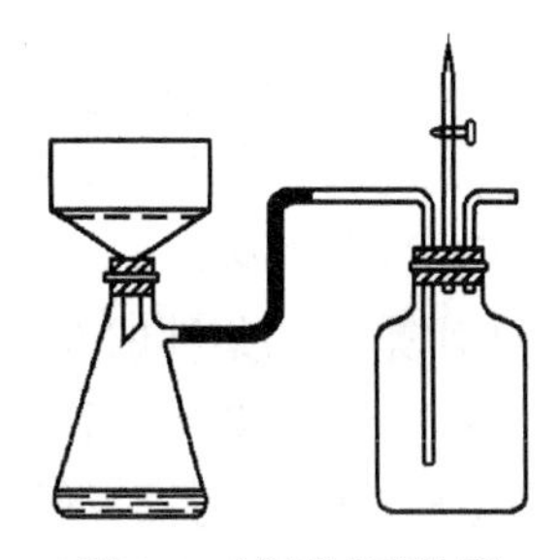
图 2.3　减压过滤装置

必须注意滤纸的直径应略小于布氏漏斗的直径，能覆盖布氏漏斗所有瓷孔。滤纸被蒸馏水浸湿后，负压下滤纸牢固粘附并完全覆盖漏斗所有瓷孔，防止晶体从边缘溢出。打开水泵，转移样品，得滤饼。用玻璃棒或刮刀小心地将晶体松开（不要松开滤纸），然后在滤饼上加入少量溶剂清洗滤饼，确保所有晶体都完全浸入溶剂中，然后再次抽吸以去除溶剂。重复两次，得到不含母液的滤饼。

3. 热过滤

在溶液冷却前，须从沸腾或热溶液中去除不溶性杂质或活性炭，这一过程称为热过滤，通常包括常压过滤和减压过滤。

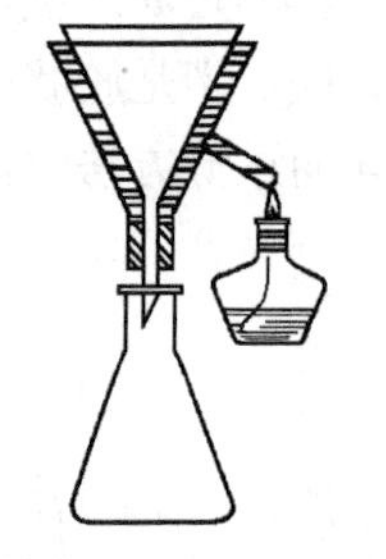

图 2.4　保温漏斗装置

如图 2.4 所示，将短茎漏斗插入铜制外套中，加热侧臂导管中的水，以保持漏斗温热。如果实验室没有保温漏斗，也可用预热的短柄漏斗代替（注意：预热后应立即进行过滤）。漏斗中放入折叠滤纸，使用前先用少量热溶剂润湿滤纸，以免干燥的滤纸吸附溶剂使溶液浓缩而析出晶体，然后在玻璃棒的导引下，将沸腾的溶液倒入滤纸中，最后用表面皿盖住漏斗，以减少热溶剂的蒸发。检查滤纸上或漏斗中是否有结晶析出，如有，加入少量沸腾溶剂直到晶体溶解。

进行热过滤操作要求准备充分，动作迅速。

2.4　萃取与洗涤

萃取是纯化或分离有机化合物的常用操作之一，可分为液-液萃取、液-固萃取。极性（或非极性）相似的液体互溶，称为“相似相溶”。当一种溶剂与另一种不混溶的溶剂振摇，两相再次分层时，达到了分配平衡，在每一种溶剂中的溶质浓度比为常数，称为分配系数 K。

2.4.1　液-液萃取

液体的萃取（或洗涤）通常是在分液漏斗中进行。常用的分液漏斗有球形、锥形和梨形三种。在有机化学实验中，分液漏斗主要用于：①分离两种不互溶且不反应的液体；②从溶液中萃取某种成分；③用水或碱或酸溶液洗涤某种产品；④用来滴加某种试剂（代替滴液漏斗）。

分液漏斗是一种玻璃实验器皿，其主要结构包括斗体、盖在斗体上口的塞子和在斗体下口具两通结构的活塞。将分液漏斗洗净，加水检漏，确认不漏水后，关好活塞。将分液漏斗置于固定在铁架台上的铁圈中，把待萃取混合液从上口倒入分液漏斗，盖好上口塞。多次翻转漏斗，小心振荡，使萃取剂和待萃取混合液充分接触，如图 2.5 所示。振荡过程中，要不时将漏斗尾部向上倾斜并打开活塞，以排出因振荡而产生的气体。振荡、放气操作重复数次后，将分液漏斗再置于铁圈中，静置分层。当两相分清后，先打开分液漏斗上口塞，然后打开活塞，使下层液体经活塞孔从漏斗下口慢慢放出，上层液体自漏斗上口倒出。这样，萃取剂溶解被萃取物质从原混合物中分离出来。通常萃取三次即可。

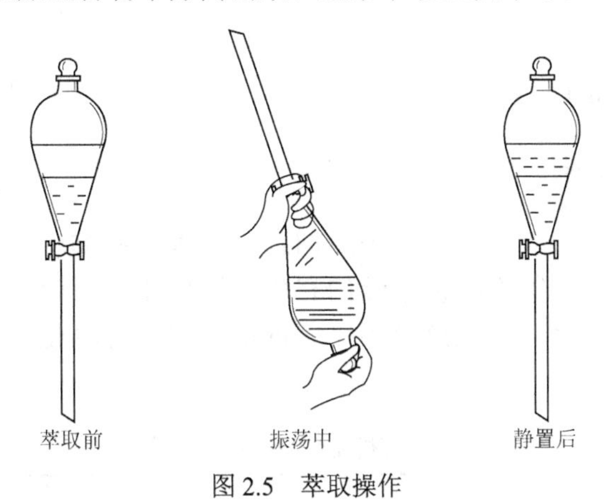

图 2.5　萃取操作

2.4.2　液-固萃取

自固体中萃取化合物，通常采用脂肪提取器（如索氏提取器），如图 2.6 所示。索氏提取器是利用溶剂回流和虹吸原理，使固体物质每一次被纯的溶剂所萃取，因而效率较高。为增加液体浸溶的面积，萃取前应先将物质研细，用滤纸套包好置于提取器中，提取器下端接盛有萃取剂的烧瓶，上端接冷凝管，当溶剂沸腾时，冷凝下来的溶剂滴入提取器中，待液面超过虹吸管上端后，即虹吸流回烧瓶，因而萃取出溶于溶剂的部分物质。这样利用溶剂回流和虹吸作用，使固体中的可溶物质富集到烧瓶中，提取液浓缩后，将所得固体进一步提纯。

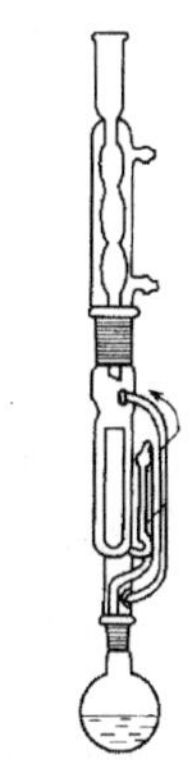

图 2.6　索氏提取器

2.5　干燥与干燥剂

有机化合物在进行波谱分析、定性或定量化学分析之前，以及固体有机物在测定熔点前都必须完全干燥，否则会影响结果的准确性；液体有机物在蒸馏前也需要进行干燥，以除去其中的水分，否则前馏分较多，测得的沸点不准确。此外，许多有机化学反应需要在“绝对”无水条件下进行，既要干燥原料、溶剂和容器，还要在反应进行时隔绝空气中的湿气。所以在有机化学实验中，试剂和产品的干燥是非常普遍且十分重要的基本操作。

干燥的方法主要有物理方法和化学方法两类。物理方法主要有吸附、分子筛脱水等；化学方法则是用干燥剂除水，根据除水原理的不同又可分为与水结合生成水合物（如氯化钙、硫酸镁和硫酸钠等）和与水发生化学反应（如五氧化二磷、氧化钙等）两种。

2.5.1　液体有机物的干燥

1. 干燥剂的选择

液体有机物的干燥通常是将干燥剂直接加入有机物中，因此选择干燥剂时要考虑以下因素：①不与被干燥的有机物发生化学反应；②不能溶于该有机物中；③吸水量大、干燥速度快；④价格便宜。有机化合物的常用干燥剂见表 2.1。

表 2.1　有机化合物的常用干燥剂

有机化合物	干燥剂
烃	氯化钙，金属钠
卤代烃	氯化钙，硫酸镁，硫酸钠
醇	碳酸钾，硫酸镁，硫酸钠，氧化钙
醚	氯化钙，金属钠
醛	硫酸镁，硫酸钠
酮	碳酸钾，氯化钙（用于干燥高级酮）
酯	硫酸镁，硫酸钠，氯化钙，碳酸钾
硝基化合物	氯化钙，硫酸镁，硫酸钠
有机酸，酚	硫酸镁，硫酸钠
胺	氢氧化钠，氢氧化钾，碳酸钾

2. 干燥剂的用量

干燥剂的用量取决于体系的含水量、干燥剂的容量和要干燥的液体量。切记，在干燥液体时使用过多的干燥剂会因干燥剂吸附而造成产品的损失，同时，被干燥溶液的机械损失也会变大。如果体系最初含水量较大，而必须加入超量的干燥剂以达到凝集点时，应向被干燥体系添加更多的有机溶剂，以尽量减少产品的损失。

通常，干燥 25 mL 有机溶液，需加入 0.5～1 g 粉末或粒状无水干燥剂。

3. 操作方法

干燥时选用的干燥剂颗粒不能太大，而粉状干燥剂在干燥过程中易呈泥浆状，导致分离困难，应慎重选择。实际操作时，通常先加入少量干燥剂，充分振荡，静置，如出现干燥剂附着器壁，相互黏结，摇动不易流动等，则说明干燥剂用量不足，应再补充干燥剂，直至新添加的干燥剂不结块，不粘壁，摇动时能自由流动，则说明所加干燥剂用量合适。如加入干燥剂后出现水相，必须用吸管将水吸出，然后再添加新的干燥剂。

干燥前液体有机物呈浑浊状，干燥后变澄清，这可简单地作为水分基本除尽的标志。需要注意的是，温度越低，干燥剂的干燥效果越好，因此干燥应在室温下进行，而干燥后液体有机物需要蒸馏时，必须将干燥剂和液体过滤分离。

2.5.2 固体有机物的干燥

干燥固体有机化合物，主要是为了除去残留在固体中的少量低沸点溶剂，如水、乙醚、乙醇、丙酮、苯等。常用的干燥方法有如下几种。

1. 风干

固体在空气中自然晾干是最简便、最经济的干燥方法。该方法适用于在空气中稳定、不易分解、不易吸潮的固体。操作时，将待干燥固体置于干燥的表面皿或滤纸上，摊开成薄薄的一层，再用另一张滤纸覆盖起来，放在空气中慢慢晾干。

2. 烘干

对于热稳定性好、熔点较高的固体有机物，可将待干燥固体置于表面皿或蒸发皿中，放在水浴上、砂浴上或两层隔开的石棉网上烘干，也可放在恒温烘箱中或红外灯下烘干。操作时，要注意防止过热，加热的温度切记不能超过该固体的熔点，以免固体变色或分解。

3. 干燥器干燥

对于具有高吸湿性和高温分解或升华特性的固体有机化合物，可以在干燥器中干燥。干燥器可分为普通干燥器、真空干燥器和真空恒温干燥器。常用干燥器类型和特点见表 2.2。

表 2.2　常用干燥器类型和特点

类型	特点简介
普通干燥器	盖与缸身之间的平面经过磨砂处理，在磨砂处涂以真空脂使其密闭。缸中有多孔瓷板，瓷板下面放置干燥剂，上面放置盛有待干燥固体的表面皿等
真空干燥器	干燥效率较普通干燥器好。真空干燥器上有玻璃活塞，用以抽真空，活塞下端呈弯钩状，口向上，防止通向大气时因空气流速过快将固体吹散。为防止干燥器在负压下爆裂，干燥器周围应以金属网或防爆布围住。解除干燥器内真空时，应缓慢打开活塞，以免空气流速过快
真空恒温干燥器	也称干燥枪，但仅适用于少量样品的干燥。将盛有样品的小瓷舟置于夹层内，连接上盛有五氧化二磷的曲颈瓶，然后减压至最高真空度时，停止抽气，关闭活塞后加热溶剂（溶剂的沸点应低于待干燥样品的熔点），利用蒸气加热夹层的外层，从而使样品在恒定的温度下被干燥

上述干燥器使用的干燥剂应根据样品所含的溶剂来选择。例如，五氧化二磷吸收水；生石灰吸收水和酸；无水氯化钙吸收水和醇；氢氧化钠吸收水和酸；石蜡片吸收乙醚、氯仿、四氯化碳、苯等有机溶剂。

2.6　蒸馏、分馏与回流

2.6.1　蒸馏

1. 原理

蒸馏是加热物质至沸，使其气化，再冷凝蒸气，并于另一容器中收集冷凝液的操作过程。它是分离和提纯液态有机化合物最常用的方法之一。应用这一方法不仅可以把挥发性物质与不挥发性物质分离，而且可以把沸点不同的液体混合物分离。但要注意由于普通蒸馏是利用液态化合物的沸点差异进行分离，所以只有当混合液体的沸点有显著不同时（至少相差 30℃以上），普通蒸馏才能将其有效分离。一个纯的液态化合物在一定压力下具有固定的沸点，所以蒸馏法还可以用于测定物质的沸点，检验物质的纯度。

2. 操作

蒸馏装置主要由气化装置、冷凝装置和接收装置三部分组成，如图 1.3 所示。蒸馏烧瓶大小应由待蒸馏液体的体积来决定，通常液体的体积应占蒸馏烧瓶容量的 1/3～2/3。加入 1～2 粒沸石。确认装置安装牢固。以水作冷却剂时，冷凝水从冷凝管套管的下端流进，从其上端流出，且上端出水口应向上，以保证套管中充满水，缓慢通入冷却水后加热。记录第一滴馏出液滴入接收瓶时的温度并接收沸点较低的前馏分。调节加热速度，使馏出液的蒸出速度以每秒 1～2 滴为宜。当温度升至所需沸点范围并恒定时，更换另一接收瓶收集，并记录此时的温度范围，即馏分的沸点范围。当烧瓶中仅残留少量液体时，停止蒸馏，移去热源，关闭冷却水。

2.6.2 分馏

1. 原理

分馏是采用分馏柱达到分离和提纯目的的方法。这种技术可以有效地分离沸点差别不大、用蒸馏方法无法分离的液体混合物。

分馏是利用分馏柱将多次气化—冷凝过程在一次操作中完成的方法。混合液受热沸腾后，当混合蒸气沿分馏柱上升时，由于分馏柱外空气的冷却作用，部分蒸气被冷凝。冷凝液在下降途中与上升的蒸气接触，二者进行热交换，蒸气中高沸点组分被冷凝，低沸点组分仍以蒸气形态上升。因此，上升蒸气中低沸点组分含量增多，而下降的冷凝液中高沸点组分增多。如此经过多次气-液两相间的热交换，就相当于连续多次的普通蒸馏过程，以致低沸点组分蒸气不断上升而被蒸馏出来，而高沸点组分则不断流向烧瓶中，从而达到分离的目的。

2. 操作

分馏装置由圆底烧瓶、分馏柱、冷凝管及接收器四部分组成，如图 1.4 所示。

与蒸馏操作相似，先将不超过烧瓶容量 2/3 的待分馏液体加入圆底烧瓶中，放入 1～2 粒沸石，确认装置安装牢固。缓慢通入冷却水后加热。调节加热速度，收集所需温度范围的馏分。当烧瓶中仅残留少量液体时，停止蒸馏，移去热源，关闭冷却水。

2.6.3 回流

许多有机化学反应需要使反应物在较长时间内保持沸腾才能完成。为了防止反应物以蒸气逸出，常用回流冷凝装置，使蒸气不断地在冷凝管内冷凝成液体，返回反应器中。

回流是一种在长时间操作过程中用来冷却混合液的技术。回流装置主要由气化装置、冷凝装置和干燥装置三部分组成，包括圆底烧瓶、冷凝器和干燥管（图 1.5）。

为了防止空气中的水蒸气进入反应容器内或吸收反应中放出的有毒气体，可在冷凝管上口连接 $CaCl_2$ 干燥管或气体吸收装置。进行回流操作时，也要控制加热，蒸气上升的高度一般以不超过冷凝管的 1/3 为宜。

2.7 升 华

升华是一种纯化微量固体物质常用且有效的方法。升华是固态有机物不经过液态直接变成气态的相变过程。通过升华提纯物质，则该化合物必须具有较高的蒸气压，并且杂质必须具有比被纯化化合物较低的蒸气压。

在实验室中，升华常被用于以下有机物的纯化：①化合物可不熔化就蒸发；②化合物稳定，且蒸发时不分解气化；③化合物蒸气可冷凝回固体；④杂质不升华。

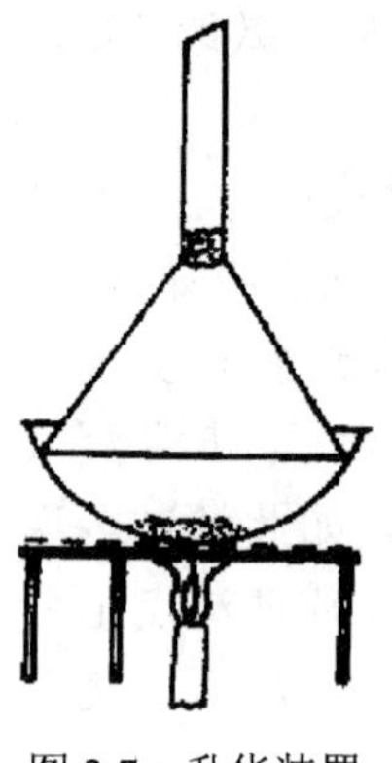

图 2.7　升华装置

如图 2.7 所示为一种常见的升华装置：首先将待纯化有机物置于升华室的底部；然后用油浴、砂浴或小火焰将样品加热到低于固体熔点的温度；被纯化有机物固体蒸发，气相转移到玻璃漏斗的表面，冷凝形成纯固体。

升华还可用于固体化合物中水和其他低沸点溶剂的去除。常见实例如冷冻干燥，在冷冻干燥过程中，溶液被冻结，压力降低，固体溶剂升华，产生的蒸气不断被排除，进而实现固体干燥。例如，从咖啡、茶和食物中除去水以减少总重量并防止变质；再如，对热敏感的蛋白质、酶和核酸可以通过减压升华除去水，进而实现纯化、回收。

2.8　色　谱　法

色谱法是一种重要的分离分析方法，常用于混合物纯化分离。

国际纯粹与应用化学联合会（IUPAC）将色谱定义为物理分离方法，其中待分离的组分分布在固定相和流动相两相之间，流动相以一定的方向移动并通过固定相，被分离的物质在分子间力作用下被吸引到固定相，吸引力越强，它们通过流动相迁移越慢，停留在固定相中时间越长，混合物分离结果依据不同化合物的迁移速率不同。

所有色谱方法都具有相同的分离原理。有机化学实验中最常用的是液-固色谱，它有两种常见类型：薄层色谱（TLC）和柱色谱（CC）。

2.8.1　薄层色谱法

薄层色谱是有机化学实验中最常用的液-固色谱形式之一，是一种简单、快速、经济的微量有机物分离分析技术，其最常见的用途是：①鉴别混合物中的各组分；②筛选、确定柱色谱分离混合物的最佳条件；③监测反应或柱色谱分离进程。

1. 薄层色谱板

薄层色谱板由玻璃、金属或塑料等固体载体组成，固体表面涂覆有薄层吸附剂作为固定相。

硅胶（$SiO_2 \cdot xH_2O$）是常用的色谱固相吸附剂。市售氧化铝（Al_2O_3）有三种类型：中性、酸性和碱性，酸性和碱性氧化铝分别用于分离碱性化合物和酸性化合物，中性氧化铝是薄层色谱吸附剂最常见的形式。不同化合物与固定相相互作用强度不同，但有一个普遍的共性是：化合物的极性越强，它与硅胶或氧化铝的结合程度越大。

2. 样品制备

样品必须溶解在挥发性有机溶剂中，浓度为 1%～2%的稀溶液效果最好。通常使用分析纯无水丙酮或乙酸乙酯。如果溶解固体，在 1 mL 溶剂中溶解 10～20 mg；如果溶解非挥发性液体，在 1 mL 溶剂中大约溶解 10 μL。

3. 点板

用微量毛细管吸取已制备的样品溶液，垂直轻轻接触薄层板起点线（距薄层板底边约 1 cm），保持点样点小而集中。点样的关键是样品不宜太多，否则易超载，导致拖尾和分离效果较差。

4. 展开

将已点好样的薄层板放入大小适宜并已装好展开剂的封闭展缸中展开。

展开前，可多次利用薄层色谱板分析选择合适的展开剂。为了确保展开分离效果，展缸中必须用展开剂蒸气饱和，以防止展开剂在薄层板爬升时溶剂蒸发。在薄层色谱板正在展开时，不要提起或以其他方式干扰展开。薄层色谱板的展开通常需要 5～10 min。当展开剂前沿距薄层色谱板顶端 1～1.5 cm 时，用镊子从展缸中取出薄层色谱板，用铅笔标记溶剂前沿。当薄层色谱板上展开剂挥发完全后即可进行显色、分析。展开过程如图 2.8 所示。

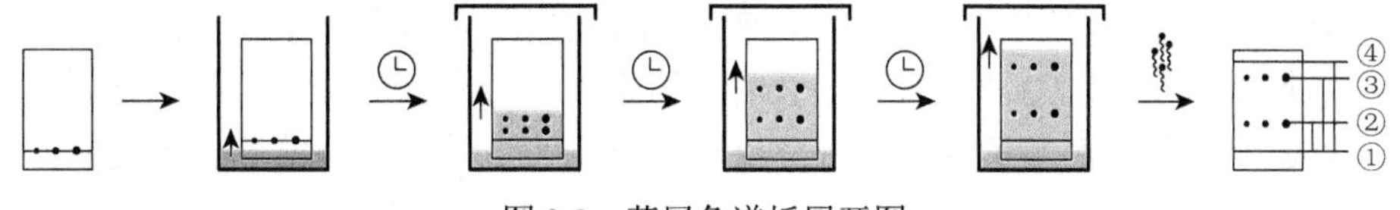

图 2.8　薄层色谱板展开图

5. 显色

有色化合物的色谱分离通常可以直接在薄层色谱板上看到，但无色化合物需要间接方法观测。最简单的显色技术包括使用含有荧光指示剂的吸附剂，在暗箱中以短波长（254 nm）紫外灯照亮色谱板，可见样品荧光点，用铅笔勾勒出每个斑点，以便对色谱板进行分析。

6. *薄层色谱分析*

薄层色谱分析包括确定在薄层色谱板上每种化合物移动距离（在测量移动距离时必须从同一起点出发）与溶剂移动距离的比值，这一距离比称为 R_f（保留因子），它是化合物的特征物理参数（图 2.9）。在一组恒定的实验条件下，每种化合物的 R_f 值取决于化合物自身结构、所用吸附剂和流动相。当进行色谱分析时，应在实验笔记上记录每种物质的 R_f 值和实验条件。

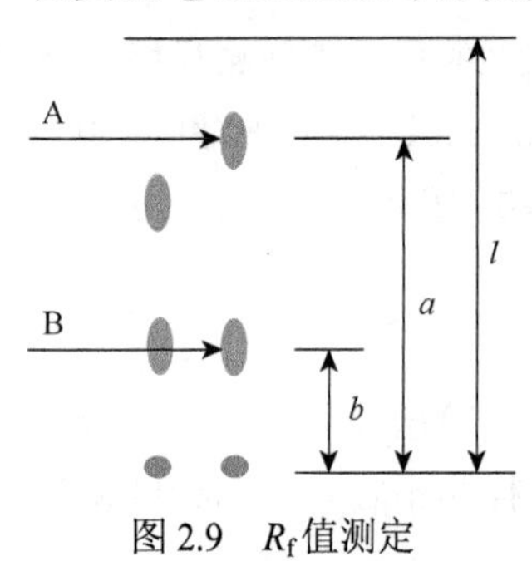

图 2.9　R_f 值测定

2.8.2　柱色谱

柱色谱属于液相色谱，与高效液相色谱（HPLC）法相似，是有机化学实验中重要的色谱方法。柱色谱分离原理与薄层色谱法相同。

柱色谱通常用于分离低挥发性化合物，而气相色谱（GC）只适用于分离挥发性混合物。与薄层色谱和气相色谱不同，液相色谱分离样品量可从高效液相色谱的几微克到柱色谱的 10 g 或更多。

1. *洗脱液*

洗脱溶剂的选择是柱色谱分离的关键。如果使用的溶剂极性大，则洗脱速度快，产物与杂质之间或混合物的组分之间几乎没有分离；如果溶剂是非极性或极性小，则组分将保留在色谱柱中。因此，应利用薄层色谱技术，确定最合适的溶剂（或溶剂混合物），使观察到的各组分斑点尽可能分开。

2. *固定相*

吸附剂（硅胶）的用量取决于待纯化样品量和薄层色谱上各组分的 R_f 值。通常，硅胶与样品量比为 50∶1。如吸附剂用量太少，则色谱柱会过载，分离效果差；反之，如吸附剂用量太多，柱色谱分离时间将更长，需要的洗脱剂更多，分离效率低。

3. *色谱柱*

通常 10～20 cm 高度的硅胶分离效果较好，吸附剂高度与色谱柱内径比为 8∶1 或 10∶1 为宜。因此，柱色谱法中通常选用柱径 1.5～2.5 cm 色谱柱。色谱柱被固定在实验室台架上，使用玻璃漏斗将固定相和洗脱液引入。

4. 加样

调整色谱柱内洗脱液液面仅达到硅胶上限。如果样品溶于洗脱液中，则用滴管将少量已溶解的样品液转移到色谱柱内硅胶上，避免溅到柱壁；如果样品不易溶于洗脱液中，则将样品溶解于硅胶混合拌样，待溶剂蒸发后，将已干燥硅胶均匀加入色谱柱中，同样避免硅胶挂侧壁。

5. 洗脱

加入洗脱剂洗脱，柱色谱分离样品时洗脱流速应调到最佳，以便实现有效分离。洗脱液流速如果太慢，样品将过度扩散到色谱柱中，分离时间延长，分离效率降低；洗脱液流速如果太快，则样品与柱内固定相相互作用时间短，分离效果差。柱色谱分离过程如图 2.10 所示。

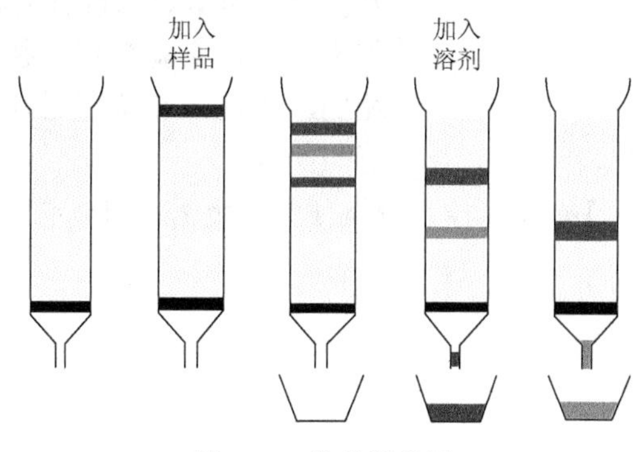

图 2.10　柱色谱装置

6. 收集洗脱液并纯化产品

将洗脱液分别收集在试管中，通过 TLC 确定和找出已分离或提纯的各组分。一旦确定相应产物后，相同洗脱液集中在圆底烧瓶中，通过旋转蒸发器减压除去溶剂，获得纯组分。

2.8.3　纸色谱

纸色谱（PC）属于分配色谱，是以滤纸作载体，使样品溶液在滤纸上展开达到分离的目的。纸色谱是用特制的滤纸作为惰性载体，以吸附在滤纸上的水或有机溶剂为固定相，流动相则是含有一定比例水的有机溶剂，通常称为展开剂。在滤纸的一定部位点上样品，当有机相沿滤纸流动经过原点时，即在滤纸上的水与流动相间连续发生多次分配，在流动相中具有较大溶解度的物质随溶剂移动的速度较快，而在水中溶解度较大的物质随溶剂移动的速度较慢，这样便达到混合物分离的目的。

纸色谱的操作过程与薄层色谱类似。在滤纸一端 2～3 cm 处用铅笔划好起始线，将被分离的样品溶液用毛细管点在起始线上，待溶剂挥发完后，将滤纸的另一端悬挂在层析缸的玻璃钩上，使滤纸下端与展开剂接触，展开剂由于毛细作用沿滤纸条上升，如图 2.11 所示。当展开剂前沿接近滤纸上端时，将滤纸条取出，记下溶剂的前沿位置，晾干。

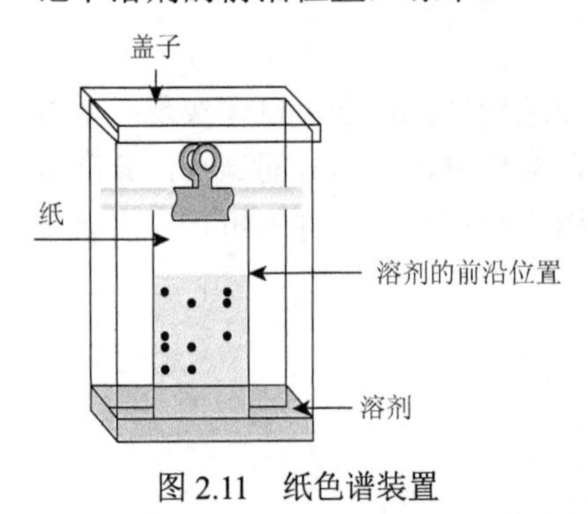

图 2.11　纸色谱装置

若被分离的各组分是有色的，滤纸条上就有各种颜色的斑点显出，否则要用紫外光或显色剂显色。

计算化合物的 R_f 值。R_f 值与被分离化合物的结构、固定相与流动相的性质、温度以及滤纸的质量等因素有关。由于影响 R_f 值的因素很多，实验数据往往与文献记载不完全相同。因此，在鉴定时通常采用标准样品作对照比较。

纸色谱法的优点是操作简单，价格便宜，色谱图可以长期保存。缺点是展开速度较慢。

Chapter 2 Basic Techniques of Organic Chemistry Experiments

2.1 Washing and Drying of Glassware

2.1.1 Washing of glassware

For the successful experiments and proper maintenance, laboratory glassware must be cleaned and dried after the end of each experiment so that dirt or chemical residues do not interfere with the completion of each new experience.

There are many washing methods for glassware. The appropriate washing methods should be selected according to the experimental requirements, the nature of the dirt and the degree of contamination.

Most chemical residues can be removed by washing the glassware using a brush, special laboratory soap and water.

Chromic acid lotion, which is made from concentrated sulfuric acid and chromic anhydride or potassium dichromate, is often an effective cleaning agent, but because it is a strong oxidizing acid, it must be used with great care. After the glassware is clean, pour the chromic acid solution into a specially designated bottle, not into the sink, lest the sewer be corroded.

Another powerful cleaning solution is organic solvents, such as acetone, which is miscible with water. It dissolves most organic residues and thus is commonly used to clean glassware.

2.1.2 Drying of glassware

(1) Air Drying. For non-urgent instruments, the instrument can be inserted upside down in the laboratory drying rack to dry.

(2) Blow dry. The instrument is inverted to control the dehydration, and the outer wall is wiped dry. The residual moisture in the instrument is driven out by the hot air of the hairdryer.

(3) Oven drying. Wet glassware can be dried by heating it in an oven at 120℃ for 20 min. Remove the dried glassware from the oven with tongs and allow it to cool to room temperature before using it for a reaction.

（4）Drying with organic solvents. A small amount of organic solvents（such as acetone，ethylalcohol，etc.）are added to the washing instrument，and the instrument is rotated to mix the water in the instrument with organic solvents and pour out the mixed liquid（recovery）. The instrument dries rapidly.

It must be noted that in some chemical experiments，the instrument do not need to be dried，such as gauges，containers，etc. Wash the residual water droplets with a small amount of solution two or three times before use. Measuring containers with scales can not be dried by heating method，otherwise the accuracy of the instrument will be affected. If drying is needed，the method of air drying or cold air drying can be used.

2.2 Heating and Cooling

2.2.1 Heating

At room temperature，some organic reactions are difficult to carry out or the reaction rate is very slow. In order to improve the reaction rate，the condition of heating is generally required，for example the processes of distillation and sublimation require heating. Several heating sources，such as gas burner，heating mantle and electric hot plate，are commonly used in the organic chemistry lab. Usually，heating is not carried out straightforwardly. To avoid possible safety problems from straightforward heating，the following indirect heating methods are generally used according to the specific circumstances.

1. Water bath

In the case of solutions of flammable liquids having a boiling point below 100℃，the electrically-heated water bath kettle provided with a constant-level device must be used（see Figure 2.1），sometimes，the beaker with larger size is often used instead. If the heating temperature is below 90℃，the vessel can be immersed into the water bath directly. The temperature of about 95℃ can be achieved by using boiling water bath or steam bath. In the operation of steam bath，the vessel is suspended above the water，which is heated by the vapor. Due to the evaporation of the water，a certain amount of hot water should be added during the operation，so that the surface of water bath is kept higher than that of the solution inside.

Figure 2.1 Water bath apparatus

2. Oil bath

For the required temperature in the range of 100-250℃, the oil bath is generally used. The temperature can be easily controlled in a certain range, and the reactant in the vessel is heated evenly. The reaction temperature is generally lower than that of the oil bath about 20℃. It is same to water bath that the surface of the oil bath should be kept higher than that of the solution inside.

The operators must take particular care to prevent fire when using the oil bath. The heating should be immediately stopped when the oil is fuming severely. Once catching fire, the operators ought to turn off the power or gas burner firstly, and then remove surrounding combustibles, finally cover the oil bath with asbestos. To avoid excessive heating, a thermometer should always be placed in the bath. Introducing water into the bath is also not permitted, which may splatter from the hot oil. Flasks should be allowed to drain for several minutes above the bath and then wiped with a rag when removed from an oil bath.

The highest temperature that oil bath can reach is dependent upon what kind of oil is used. Several heat conducting oils, commonly used in the organic chemistry lab, are presented as follows:

(1) Glycerol is satisfactory up to 140-150℃. Above these temperatures, decomposition is usually excessive and the odor of the vapors is unpleasant. Due to its high hydrophilicity glycerol should be heated to remove the absorbed moisture before use when being laid aside for a long time.

(2) Plant oil, such as rapeseed oil, castor oil and peanut oil, for temperatures up to about 220℃, is recommended. One percent of hydroquinone or other antioxidant is usually added to improve the thermal stability of the oil. It should be noted that plant

oil easily decomposes at high temperature and combustion occurs when reaching its flash point.

（3）Paraffin can be heated to about 200℃. It is convenient for storage because the hot oil will turn into a solid when cooling down to room temperature. Therefore, paraffin should be taken out from the bath immediately after use.

（4）Liquid paraffin can be heated up to about 200℃. It is of high thermal stability, but easier to burn at high temperature.

（5）Silicone fluid is probably the best liquid for oil bath but somewhat expensive for general use. It may be heated high up to 250℃ without appreciable decomposition and discoloration.

3. Sand bath

When the required heating temperature is much higher than those listed above, one can often use a sand bath. Its highest operating temperature is up to 350℃. The clean and dry sand is placed in sand bath iron box, in which the vessel is half buried. The disadvantage of the sand bath is the poor heat conductivity, but fast heat dissipation, resulting in uneven distribution of the temperature. Hence the sand at the bottom of the sand bath iron box should be slightly thinner, but slightly thicker around the vessel. A thermometer should always be inserted in the bath to detect the temperature of which the mercury bulb should be close to the vessel.

4. Air bath

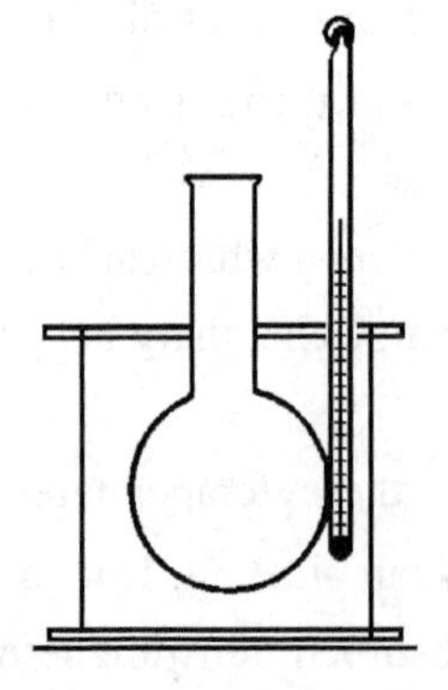

Figure 2.2 Air bath apparatus

Air bath is a very cheap and convenient method of effecting heating of small distillation flasks, which is satisfactory for the liquids having a boiling point above 80℃. When using gas burner as heating source, it is essential to use a heat resistant bench mat, which is away from the flask about 1 cm. This is the simplest air bath (see Figure 2.2). However, this method easily results in fluctuations at the level of heating due to air draughts. It is not suitable for the reflux of liquids with low boiling point or distillation under reduced pressure.

5. Heating mantle

The heating mantle, or isothermal jacket, is the most commonly used as a piece of laboratory equipment to apply heat to containers. They consist of a heating element enclosed within knitted glass-fiber, which makes it much safer. In contrast to other heating devices, such as electric heating plate or gas burner, glassware containers may be placed in direct contact with the heating mantle without substantially increasing the risk of the glassware shattering, because the heating element of a heating mantle is insulated from the container so as to prevent excessive temperature gradients. Because the heating mantle is not heated by open fire, it can be used to heat and distill flammable organic compounds or compounds with higher boiling point. Heating mantle is satisfactory from room temperature to about 200℃ and is applicable to a wide range of heating temperature.

2.2.2 Cooling

Exothermic reactions are usually accompanied by a large heat generation, which increases the reaction temperature rapidly. This tends to cause evaporation of the reactions, even an explosion. It is often necessary to obtain low temperature for these reactions by immersing the reaction vessel in a cooling bath. This method is also used in the processes of separation and purification reactants. The coolants should be selected according to the low temperature to be achieved and the quantity of heat to be taken away.

(1) Water. Water is the most commonly used coolant because of its lowest cost and high thermal capacity. It should be noted that the cooling efficiency greatly changes with different seasons. Moreover, water is often used in reflux and distillation to cool the hot gas or liquid.

(2) Ice-water mixture. Finely crushed ice is used for maintaining the temperature at 0-5℃. It is usually best to use a slush of crushed ice with water to provide sufficient contact with the vessel to be cooled. If the water does not interfere with the reaction, crushed ice can be directly put into the reaction mixture to maintain low temperature effectively.

(3) Ice-salt mixture. For temperatures below 0℃, the commonest freezing mixture is a mixture of crushed ice and inorganic salt in certain ratio. A mixture of sodium chloride and ice is most commonly used in the lab, which can produce a temperature of –21℃ to –5℃.

(4) Dry ice. Dry ice is employed to obtain the temperature below –60℃. If small pieces of dry ice are carefully added to either ethanol or acetone until the lumps of dry ice no longer evaporates vigorously, the temperatures attained are low to –78℃. In

order to keep the freezing mixture for hours or overnight，it should be prepared in a Dewar flask.

（5）Liquid nitrogen. The attainment of temperatures lower than –100℃ requires the use of baths employing liquid nitrogen，which can be as cold as –196℃. It should also be stored in a Dewar flask to keep the cooling efficiency.

It should be noted that the mercury thermometer is forbidden to use when the temperature is below –38℃，because the freezing point of mercury is –38.9℃. It is preferably to choose the low temperature thermometer based on ethanol，toluene or *n*-pentane with a little pigment.

2.3 Recrystallization and Filtration

2.3.1 Recrystallization

Recrystallization，a frequently used method and a valuable technique to purify organic solids，is based on the fact that the solubility of an organic compound in a specific solvent will increase significantly as the solvent is heated. When an impure organic solid is dissolved in such a heated solvent and then the solution is cooled to room temperature or below，the solid will usually recrystallize from solution in a pure form. This allows the solid to be isolated in a pure form if the impurities in the solid either fail to dissolve at the elevated temperature or remain dissolved when the solution is cooled. Thus，as the mixture cools，the impurities tend to remain in solution while the highly concentrated product crystallizes.

The general process of recrystallization can be broken into six steps：choosing the solvent and dissolving the solute（i.e. preparing of a nearly saturated solution at the boiling point），decolorizing the solution，thermal filtration，crystallizing the solute collecting and washing the crystals，and drying the product.

1. Choosing the solvent and dissolving the solute

The most desirable characteristics of a solvent for recrystallization are as follows：

（1）It will not react chemically with the solute to be purified.

（2）It will dissolve the solute when the solution is hot but not when the solution is cold.

（3）It will not dissolve the impurities at all（so they will be removed from the nearly saturated solution by hot filtration），or it will dissolve them very well（so they will not crystallize along with the solute）.

(4) It will yield well-formed crystals of the solute.

(5) Non-toxic or low toxicity, easy operation.

(6) It will be inexpensive and available.

If two or more solvents appear to be equally suitable for recrystallization, the final selection will depend upon such factors as the ease of manipulation, toxicity, flammability and cost. If no information is already available in chemistry handbook or literature, it should be chosen by experimentation according to the principle of "similarity and intermiscibility".

The procedure of picking a solvent is described as follows: put a small amount of the impure solute in several small test tubes (usually about 0.2 g for each tube), and then add 0.5 mL of different solvents into the tubes, respectively. Heat the solution until all the solute dissolves, and then cool down to room temperature. The one, in which the maximum crystals are obtained, is generally considered to be the most desirable solvent. If the solute is still not completely dissolved in 3 mL of hot solvent, this solvent is unsuitable. If the solute dissolves in the hot solvent, but no crystallization occurs when the solution is cooled, even after scratching the inwall of the tube below the surface of the solution with a glass rod, this indicates that the solute to be purified has large solubility in the solvent. It should be rejected.

If the solute is found to be far too soluble in one solvent (called as the better solvent) and much too insoluble in another solvent (called as the poorer solvent), mixed solvents or "solvent pairs" may frequently be used for recrystallization with excellent results. The two solvents must, of course, be completely miscible. Solvent pairs which may often be employed include: ethanol and water, ethanol and diethyl ether, ethanol and acetone, diethyl ether and petroleum ether, benzene and petroleum ether, etc. In general, dissolve substances that are similar to it in chemical structure.

Recrystallization is generally carried out in a conical flask or a beaker. While a round flask of suitable size fitted with a reflux condenser must be employed, when using organic solvent with high flammability and volatility. Place the substance to be recrystallized in an appropriate flask, add enough solvent (less than the required quantity) to cover the solute and several boiling stones, and then heat the flask on a water bath (if the solvent boils below 90℃) or a heating mantle until the solvent boils. Stir the mixture with a glass rod to promote dissolution. Add solvent gradually, keeping the mixture at the boiling point, until all of the solute dissolve.

In recrystallization, the amount of solvent must be paid great attention to in order to obtain the product with higher purity and yield. The amount of solvent should be considered from two aspects: one is to prevent the loss of solute, which is caused by excessive solvent, the other is to consider that the solution becomes supersaturated due to the volatilization of solvents and the decrease of temperature during thermal filtration, which results in crystallization on filter paper during filtration, thus affecting the yield. Therefore, the amount of solvent can not be too much, nor too little, generally 15%-20% more than the requirement.

2. Decolorizing the solution

Occasionally an organic reaction will produce high molecular weight by-products that are deep colored. These colored impurities can be adsorbed onto the surface of activated charcoal. If too little charcoal is added, the solution will still be colored after filtration, making repetition necessary; if too much is added, it will absorb some of the product in addition to the impurities. Gradually add a small amount (1%-5% of the solute weight is sufficient) of activated charcoal to the colored solution and then boil the solution for 5-10 min. Remove the activated charcoal by filtration as described in Step 3.

3. Thermal filtration

In order to remove insoluble impurities, thermal filtration is necessary. For more details about the thermal filtration, refer to Section 2.3.2.

4. Crystallizing the solute

The saturated solution, obtained by hot filtration, is allowed to cool to room temperature. Crystallization should begin immediately. The quality of the crystals will be different under different cooling conditions. If the concentration of the solution is large and the cooling rate is high, the crystals obtained will be much smaller, and also not pure enough. If the crystallization has occurred in the process of hot filtration, the solution should be reheated to dissolve the crystals, and then stood on the bench to cool slowly without being disturbed, thus giving the large, well-formed and highly pure crystals. Once the crystallization ceases at room temperature, the flask should be placed in an ice-water bath to cool further. If no crystallization occurs after cooling, scratch the inside of the vessel with a glass rod or add a seed crystal (i.e., crystals of the same substance) to provide nucleation center, thus promoting the formation of the crystals.

5. Collecting and washing the crystals

Once the crystallization is complete, the apparatus for vacuum filtration is generally employed to separate the crystals from the mother liquor containing soluble impurities.

6. Drying the product

In order to ensure the purity of the product, the crystallization after vacuum filtration must be fully dried because there are a small amount of solvents on the surface. Different drying methods, such as natural air drying, infrared lamp drying and vacuum constant temperature drying can be used according to the crystalline properties.

After full drying, the purity of the crystallization is identified by determining the melting point. Caculate the yield of the products. If the purity does not meet the requirement, the above operations can be repeated until the melting point meets.

2.3.2 Filtration

Filtration is solid-liquid separation method by porous media (such as filter paper, filter cloth) to block the solid particles of the solid-liquid suspension. By filtering, the solid materials remains on the filter paper and the liquid filtration goes through the filter into the receiving container, the solution obtained by filter is called filtrate. Commonly, the filtering methods are atmospheric pressure filtration, vacuum filtration, thermal filtration.

Temperature, viscosity and filtration pressure of the solid-liquid mixture and the state of the material will affect the speed of filter. The greater viscosity of the solution, the slower of the filter; decompression filter is faster than atmospheric pressure; it must be destroyed by heating when it is a rubbery solid, to avoid it through filter paper. In a word, it is necessary to consider various factors when choosing different filtering methods.

1. Atmospheric pressure filtration

The method of atmospheric pressure filtration is the filter method by using common filter funnel under atmospheric pressure. The driving force of filter is gravity. So it is the most simple and general, but its speed is slow. This method is commonly used in the removal of insoluble substances. Sometimes it can be used to separate crystals from the larger particles，absorbent poor crystal products. The method is also used in analytical chemistry experiment.

In experimenting，according to the nature of precipitation to choose the type of filter paper，we will choose slow filter paper that fines crystal precipitation，we will choose fast filter paper in colloidal precipitation，and we will choose medium-speed filter paper in coarse crystal precipitation. We will select the size of the filter paper in accordance with the size of filter funnel.

2. Decompression Filtration

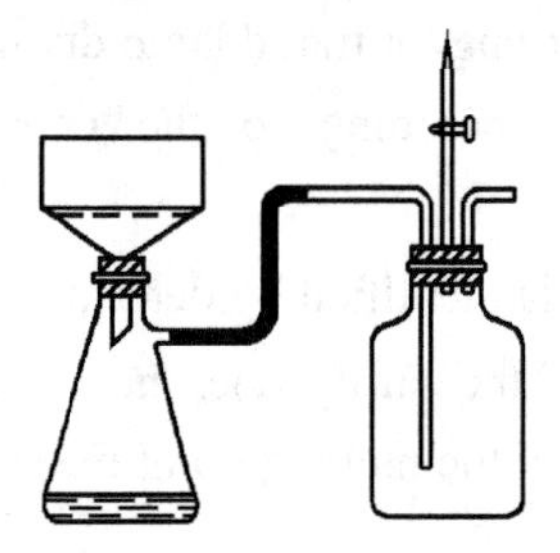

Figure 2.3 Apparatus for vacuum filtration

Decompression filtration is rapid to take away air from filter bottle by using the suction pump，so through creating the pressure difference of Buchner funnel surface and filter bottle，the filtration is speeded up. Safety bottle is assembled between the pump and suction filter bottle，to prevent the phenomenon of reverse suction because of the closing of pump. This is the apparatus of decompression filtration（see Figure 2.3）.

As shown in Figure 2.3，a porcelain Buchner funnel（the bevel of the neck of the funnel should face the side arm of the filter flask）is fitted with a rubber stopper，which is seated in the neck of a filter flask：the side arm of the filter flask is connected to a filter trap using heavy-wall rubber tubing，then to a water aspirator；the filter trap is fitted with a stopcock to regulate the pressure in the apparatus，which is essential since a sudden fall in water pressure may result in the water being sucked back and contaminating the filtrate.

You must pay attention that the filter paper should be slightly smaller than the diameter of Buchner funnel，but able to cover all porcelain holes. The filter paper is moistened with a few drops of solvent and the suction of the water aspirator applied then the filter paper should adhere firmly to，and completely cover the perforated plate of the funnel and thus prevent any crystals escaping around the edge. Open the pump after the filter paper wetted with distilled water，then transfer samples，obtain filter cake. The filter cake is then washed with fresh and cold solvent to remove the mother liquor. Add a small amount of the solvent on the filter cake and loosen the crystals carefully with a glass rod or scraper（do not loosen the filter paper）. Be sure that all the crystals are exactly immersed in the solvent, and then suck again to remove the solvent. Repeat this process two times, thus giving a filter cake without mother liquor.

3. Thermal filtration

The insoluble impurities or the activated charcoal must be removed from the boiling or hot solution before undue cooling has occurred. This process, called as thermal filtration, which commonly includes common filtration and vacuum filtration.

As shown in Figure 2.4, a short-stem funnel is inserted in a Jacket made of copper. Heating the side arm of the jacket, in which the water will be heated to keep the funnel warm. If the heat preservation funnel is not available in the lab, a preheated short-stem funnel can also be used to replace it. Note that the preheating should be immediately followed by the filtration. Filter paper is folded into a folded filter paper and wetted with a small amount of hot solvent. Grasp the vessel in a towel, and then pour the boiling solution into the paper, guided by a glass rod. Cover the funnel with a watch glass to decrease the evaporation of the hot solvent. Finally, check to see whether crystallization is occurring on the filter paper or in the funnel. If it does, add a small amount of boiling solvent until the crystals dissolve.

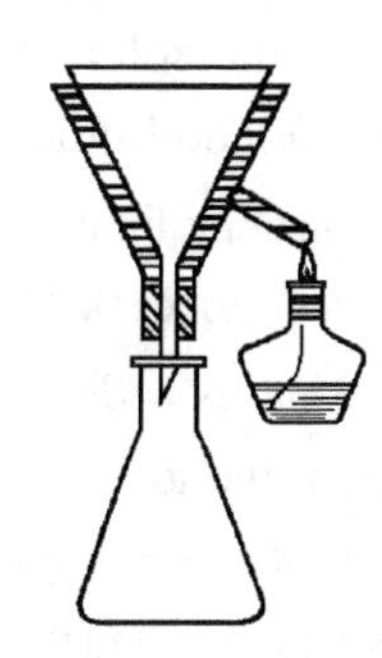

Figure 2.4 Heat preservation funnel

Full preparation and quick action are required for thermal filtration operation.

2.4 Extraction and Washing

Extraction, which includes liquid-liquid extraction and liquid-solid extraction, is an important part of the purification and separation methods. Any liquid substance with similar polarity (either polar or nonpolar) dissolves into each other, so called "like dissolves like". When a solution is shaken with a second solvent with which it is immiscible, the solute distributes itself between the two liquid phases. When the two phases have separated again into two distinct solvent layers, equilibrium will have been achieved so that the ratio concentrations of the solute in each layer define a constant, which is called the distribution coefficient K.

2.4.1 The extraction of liquid-liquid

The extracting from the liquid（or washing）is usually carried out through separatory funnel. There are three kinds of separatory funnel：spherical，subulate and pear forms. In organic chemical experiments，the separatory funnel is mainly applied to：①separate two kinds of liquid which are immiscible and do not react with each other；②extract a certain component from a solution；③wash a liquid product or the solution of a solid product with water，alkaline or acidic aqueous solution；④add liquid（replace dropping funnel）.

A separating funnel takes the shape of a cone with a hemispherical end. It has a stopper at the top and stopcock（tap）at the bottom. Check the separatory funnel and make sure that it is not leaking by adding some water in the funnel. Pour the solution to be separated in the funnel from the top. The funnel is then closed and shaken gently by inverting the funnel multiple times（see Figure 2.5）. During the shaking，the stopcock should be turned on carefully with the funnel slightly inverted to release excess vapor pressure，which is caused by quick evaporation of the organic solvent under shaking. The separatory funnel is set aside to allow for the complete separation of the two phases. The lower layer will be removed through the stopcock. The upper layer must be swilled from the top of the funnel for avoiding contamination. Generally，three times of extraction will get an effective result.

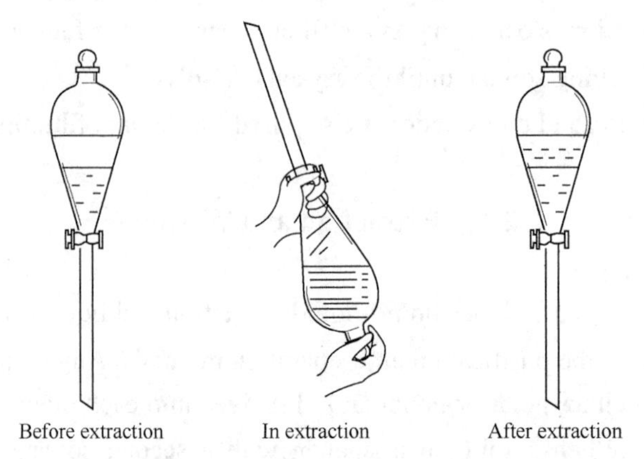

Figure 2.5　Extraction operation

2.4.2 The extraction of liquid-solid

For a solid mixture，liquid-solid extraction can be performed with a Soxhlet extractor（see Figure 2.6）. The solid is placed in a porous thimble. The extraction-solvent vapor，generated by refluxing the extraction solvent contained in the distilling pot, passes up through vertical side tube into the condenser. The liquid condensate drips onto the solid, which is extracted, the extraction solution passes through the pores of the thimble, eventually filling the center section of the Soxhlet. The siphon tube also fills with extraction solution and when the liquid level reaches the top of the tube, the siphoning action commences and the extract is remained to the distillation spot. The cycle is automatically repeated numerous times. In this manner the desired species is concentrated in the distillation pot. Equilibrium is not generally established in the system, and usually the extraction effect is very high. After the liquid-solid extraction, the solution can be treated as usual to obtain the desired product.

Figure 2.6 Soxhlet extractor

2.5 Drying and Drying Agents

Before spectral analysis, qualitative or quantitative chemical analysis, and determination of the melting point, organic compounds must be completely dried, otherwise it will affect the accuracy of the results. Liquid organic compounds also need to be dried to remove the water prior to distillation, or else much more front distillates and inaccurate melting point will be obtained. In addition, many organic reactions require the absolutely anhydrous conditions. It is necessary to dry the reactants, solvents and vessels, and the reaction should be carried out in the absence of moisture as well. Therefore, drying of the organic compounds is a very useful and important technique in organic experiments.

The common drying method can be primarily divided into physical and chemical ones. The physical drying method includes absorption, dehydration of molecular sieve, etc. The chemical drying method is to use drying agent to remove water, based on the principle of combination with water to form hydrates (e.g. calcium chloride, magnesium sulfate, sodium sulfate, etc.) and reaction with water directly (e.g. phosphorus pentoxide, calcium oxide, etc.) .

2.5.1 Drying of liquid organic compounds

1. Selection of drying agents

Liquid organic compounds are usually dried by direct contact with a solid drying agent. The selection of a drying agent will be governed by the following considerations：①it must not react with the compound to be dried；②it should be insoluble in the liquid；③it should have a rapid action and an effective drying capacity；④it should be as economical as possible. The various common drying agents are discussed in detail below. Common drying agents for each kind of organic compounds are listed in Table 2.1.

Table 2.1　Common drying agents for organic compounds

Organic Compounds	Drying Agents
hydrocarbon	$CaCl_2$, Na
halohydrocarbon	$CaCl_2$, $MgSO_4$, Na_2SO_4
alcohol	K_2CO_3, $MgSO_4$, Na_2SO_4, CaO
phenol	$CaCl_2$, Na
aldehyde	$MgSO_4$, Na_2SO_4
ketone	K_2CO_3, $CaCl_2$（for senior ketones）
ester	$MgSO_4$, Na_2SO_4, $CaCl_2$, K_2CO_3
nitro-compound	$CaCl_2$, $MgSO_4$, Na_2SO_4
organic acid，phenol	$MgSO_4$, Na_2SO_4
amine	NaOH, KOH, K_2CO_3

2. Dosage of drying agents

The amount of drying agent required depends upon the quantity of water present，the capacity of the drying agent，and the amount of liquid to be dried. Remember that the use of too much drying agent when drying a liquid can cause a loss of product by its adsorption on the drying agent，meanwhile，mechanical losses on filtration or decantation of the dried solution may also become significant.

If you have to add quite a bit of drying agent to reach the clumping point，you must have a large amount of water present initially. In this case you may wish to add more organic solvent to minimize the loss of product.

Usually，to drying 25 mL organic solution，it is required to add 0.5-1 g powder or granular anhydrous desiccant.

3. Procedure

The desiccant particles selected should not be too large，and the powdered desiccant is easy to form a slurry-like solution，resulting in separation difficulties. In practical operation one should firstly add a small amount of desiccant to the organic liquid, then swirl the solution and let it stand while observing the desiccant. If it attaches to the vessel, which is all clumped together and difficult to flow, a little more desiccant should be added until no formation of clumps for the newly added desiccant after shaking. If sufficient water is present to cause the separation of a small aqueous phase, it must be sucked out of the solution by a dropper and the solution must be treated with a fresh portion of the desiccant.

In general, the organic liquid is turbid before drying, and becomes clear after drying which can be regarded as a sign of sufficient drying. It should be noted that the lower the temperature, the better the drying efficiency. Hence the drying should be carried out at room temperature. At higher temperatures the vapor pressures above the desiccants become appreciable and the water may be returned to the liquid. The drying agents must be removed by filtration before distillation.

2.5.2 Drying of solid organic compound

Drying of solid organic compounds are mainly to remove the small amount of solvent with low boiling point in the solid, such as water, ether, ethanol, acetone and benzene, commonly used in the organic chemistry lab, are presented as follows:

1. Air dry

The organic solid is naturally dried in the air, being the most convenient and economical method for drying. This method is applicable to the drying of solid, which is stable and non-hygroscopic in the air. The solid samples should be spread out on a dry watch glass or filter paper, then covered by another filter paper and allowed to air dry.

2. Oven dry

For a solid organic compound with high thermal stability and melting point, it can be dried more quickly on a watch glass or evaporation dish, then it can be heated on a hot-water bath, in an oven or under an infrared lamp. It should be noted that the heating temperature must be lower than its melting point in order to avoid discoloration and decomposition.

3. Drying in a desiccator

For the solid organic compound with high hygroscopicity and a feature of decomposition or sublimation at high temperature，it can be dried in a desiccator. The desiccators are divided into ordinary desiccator（i.e. atmospheric pressure desiccator），vacuum desiccator and thermostatic vacuum desiccator. Types and characteristics of common drying desiccators are listed in Table 2.2.

Table 2.2 Types and characteristics of common drying desiccators

Types	Summary
ordinary desiccator	The joint between the lid and the base needs lubrication（e.g. with vacuum grease）to seal the desiccator before using it. There is a porous porcelain plate in the desiccator，below which the drying agent is placed to absorb the solvent and above which a watch glass containing the solid to be dried is placed
vacuum desiccator	The drying efficiency of the vacuum desiccator is much better than that of the ordinary desiccator. It is fitted with a glass stopcock. In this case the air inlet to the desiccator terminates in a hooked extension which serves to ensure that the air flow is directed in an even upward spread to prevent dispersal of the sample when the vacuum is released. The desiccator implosion may occur at any time when it is under vacuum，and represents a serious hazard. In use，all vacuum desiccator must be sited in an appropriately sized and totally enclosed wire-mesh desiccator cage. When releasing the vacuum，the stopcock should be slowly opened to avoid the overquick air flow
thermostatic vacuum desiccator	For smaller amounts of sample，a convenient lab vacuum oven is the so-called “drying pistol”. The drying chamber is connected to the vessel containing the drying agent（e.g. phosphorus pentoxide），which is attached to a water aspirator. The vapor from a boiling liquid in the flask rises through the jacket，surrounds the drying chamber（holding the sample in a sample boat）and returns to the flask from the condenser

The desiccant used in the above dryer should be selected according to the solvent contained in the sample. For example：phosphorus pentoxide absorbs water；lime absorbs water and acid；anhydrous calcium chloride absorbs water and alcohol；sodium hydroxide absorbs water and acid；paraffin wax absorbs organic solvents such as ether，chloroform，carbon tetrachloride and benzene.

2.6 Simple Distillation，Fractional Distillation and Reflux

2.6.1 Simple distillation

1. Principle

Simple distillation is the process of separating the components or substances from a liquid mixture by using selective boiling and condensation. Simple distillation is a technique that is used to purify a mixture of liquids. Utilizing this method，not only volatile and non-volatile substances can be separated，but also liquid mixture with different boiling points can be separated. However，be aware that simple distillation is effective only when the liquid boiling points differ greatly（rule of thumb is 30 °C）. The boiling point is a physical property of a pure organic liquid, which can be determined using the technique of simple distillation.

2. Procedure

The apparatus for simple distillation mainly consists of three parts (i.e. gasification unit, condensing unit and receiving unit). A simple distillation apparatus is set up according to Figure 1.3. The liquid to be distilled is placed in the distilling flask. The flask should be no more than two-thirds full at the start of the distillation because the contents may foam and boil over. Remember to place one or two boiling stones in the flask to promote even boiling. Ensure that all joints are tight. Turn on the water for the condenser. Only a small stream of water is needed, too much water pressure will cause the tubing to pop off. Turn on the heater and the distillation can begin. Adjust the heater until the distillate drops at a regular rate of about 1-2 drops per second. At the point at which the temperature has become the highest and most stable, collect the remaining distillate and record the temperature. This is your boiling point (never boil the distilling flask to dryness). Stop the distillation when a small amount of liquid is left in the distilling flask. Turn off the heater to allow the flask to cool more quickly, turn off the water for the condenser.

2.6.2 Fractional distillation

1. Principle

Fractional distillation is a technique used when a mixture of two liquids can not be separated well enough by using simple distillation.

Fractional distillation essentially performs these "redistillation" automatically in a portion of the fractional distillation apparatus called the fractionating column. As the mixture boils, it turns to vapor and rises up along the column. It reaches a certain point up the column and condenses. It then turns to vapor again and rises a little bit further up the column. This continues until the vapor rises all the way up the column and condenses in the condenser. At this point, the liquid has been completely purified. The distillate is the liquid with lower boiling point, while the liquid with higher boiling point remains in the distilling flask.

2. Procedure

The apparatus for fractional distillation mainly consists of four parts (i.e. round flask, fractionating column，condenser，receiver）. A fractional distillation apparatus is set up according to Figure 1.4. The procedure for fractional distillation is very similar to that of simple distillation. Once you have set up your fractional distillation apparatus，place the liquid to be distilled in the distilling flask. The flask should be no more than two-thirds full at the start of the distillation. Remember to place one or two boiling stones in the flask to promote even boiling. Ensure that all joints are tight. Turn on the water for the condenser. Turn on the heater and the distillation can begin. As always，never boil the flask to dryness. Stop the distillation when a small amount of liquid is left in the distilling flask. Turn off the heater to allow the flask to cool more quickly.

2.6.3 Reflux

In many organic chemical reactions，it is necessary keep the reactants boiling for a long time. In order to prevent the reactant from escaping by steam，a reflux condensation device is often used to make the steam continuously condense into liquid in the condensation tube and return to the reactor.

Reflux is a technique that is used to cool the mixture liquids during a long time procedure. The reflux apparatus mainly consists of three parts（i.e. gasification unit，condensing unit and drying unit），which includes round flask，condenser and drying tube（see Figure 1.5）.

In order to prevent the humidity in the air from being immersed in the reactor or to absorb the toxic gas released from the reaction，a $CaCl_2$ drying tube or a gas absorption device can be connected at the top of the condensation tube. In the reflux operation，heating should also be controlled. Generally，the height of vapor rise should not exceed one-thirds of the condensation tube.

2.7 Sublimation

Sublimation is a common and effective method for the purification of solid substances at the microscale level.

The advantage of this process is if the vapor pressure of a solid is greater than the ambient pressure at its melting point，then the solid undergoes a direct-phase transition to the gas phase without first passing through the liquid state. To be purified by sublimation，a compound must have a relatively high vapor pressure，and the impurities must have vapor pressures significantly lower than that of the compound being purified.

In the laboratory sublimation is used as a purification method for an organic compound: ①if it can vaporize without melting, ②if it is stable enough to vaporize without decomposition, ③if the vapor can be condensed back to the solid, and ④if the impurities present do not also sublime.

One common type of sublimation apparatus is shown in Figure 2.7.

In order to purify an impure substance by sublimation, it is first placed at the bottom of the sublimation chamber. The sample is then heated under reduced pressure, using an oil or sand bath or a small flame, to a temperature below the melting point of the solid. The solid will be vaporized and transferred via the vapor phase to the surface of the glass funnel, where it condenses to form a pure solid.

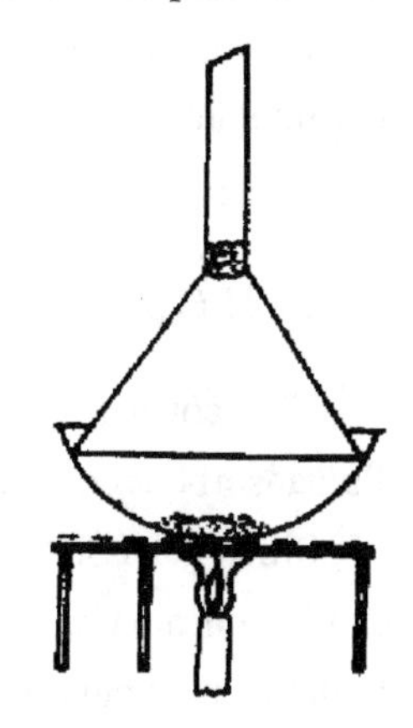

Figure 2.7 Sublimation apparatus

Sublimation is used to remove water and other solvents from solids that have lower vapor pressures than the liquid solvent. In the process known as lyophilization, the solution is frozen and the pressure is reduced. The solid solvent sublimes and the resulting vapor is removed. For example, water is removed from coffee, tea, and foods in this way to reduce their total weight and to prevent spoilage. Proteins, enzymes, and nucleic acids that are sensitive to heat can be recovered from frozen aqueous solutions by removing the water by sublimation under reduced pressure.

2.8 Chromatography

Chromatography is a separation technique frequently used for purifying substances and separating mixtures at a preparative scale.

The International Union of Pure and Applied Chemistry (IUPAC) defines chromatography as a physical method of separation in which the components to be separated are distributed between two phases: the immobile stationary phase and the mobile phase. The mobile phase moves in a definite direction and passes over the stationary phase. The substances being separated are attracted to the stationary phase by intermolecular forces; the stronger the attraction the slower they migrate through the mobile phase. Separation results from the different migration rates.

All chromatographic methods are based on the same general principle. The most commonly used in organic chemistry experiment is liquid-solid chromatography in its two types：thin layer chromatography（TLC）and column chromatography（CC）.

2.8.1 Thin layer chromatography

This is one of the most popular liquid-solid chromatography variants. It is a simple，fast，and inexpensive technique that requires only a small amount of material，and its most common uses are：①to identify the components of a mixture；②to determine the appropriate conditions for the separation of mixtures by CC；③to follow the process of a reaction or CC.

1. TLC plate

TLC consists of a solid support，such as glass，metal，or plastic with a thin layer of an adsorbent coating the solid surface，which provides the stationary phase.

Silica gel（$SiO_2 \cdot xH_2O$）is the most commonly used general-purpose adsorbent for partition chromatography of organic compounds. Aluminum oxide（Al_2O_3，also called alumina）is commercially available in three forms：neutral，acidic，and basic. Acidic and basic alumina are sometimes used to separate basic and acidic compounds respectively，but neutral alumina is the most common form of this adsorbent for TLC. The strength of the interaction varies for different compounds，but one generality can be stated：the more polar the compound，the more strongly it binds to silica gel or alumina.

2. Sample preparation

The sample must be dissolved in a volatile organic solvent，a very dilute（1%-2%）solution works best. Anhydrous reagent-grade acetone or ethyl acetate is commonly used. If you are analyzing a solid，dissolve 10-20 mg of it in 1 mL of the solvent. If you are analyzing a nonvolatile liquid，dissolve about 10 μL of it in 1 mL of the solvent.

3. Sample application

Tiny spots of the dilute sample solution are carefully applied with a micropipette near one end of the plate（1 cm）. Keeping the spots small assures the cleanest separation. It is important not to overload the plate with too much sample，which leads to large tailing spots and poor separation.

4. Development

Development of a TLC plate is carried out in a closed developing chamber containing a developing solvent.

If a developing solvent is not specified for the system you are analyzing, TLC analysis can be used to choose a suitable developing solvent. To ensure good chromatographic resolution, the developing chamber must be saturated with solvent vapors to prevent the evaporation of solvent from the TLC plate as the solvent rises up the plate. Do not lift or otherwise disturb the chamber while the TLC plate is being developed. The development of a chromatogram usually takes 5-10 min if the chamber is saturated with solvent vapor. When the solvent front is 1-1.5 cm from the top of the plate, remove it from the developing chamber with a pair of tweezers and immediately mark the adsorbent at the solvent front with a pencil. Allow the developing solvent to evaporate from the plate before visualizing the results (see Figure 2.8).

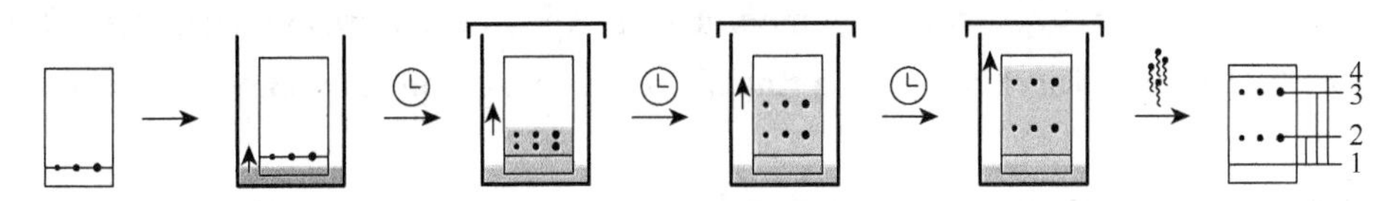

Figure 2.8 TLC chamber

5. Visualization Techniques

Chromatographic separations of colored compounds can usually be seen directly on the TLC plate, but colorless compounds require indirect methods of visualization. The simplest visualization technique involves the use of adsorbents that contain a fluorescent indicator. When the output from a short-wavelength ultraviolet lamp (254 nm) is used to illuminate the adsorbent side of the plate in a darkened room or dark box, the plate fluoresce visible light. Sometimes substances being analyzed are visible by their own fluorescence, producing a brightly glowing spot. Outline each spot with a pencil while the plate is under the UV source to give a permanent record, which will allow the analysis of your chromatogram.

6. Analysis of a TLC

After the spots on the chromatogram are visualized, you are ready to analyze the chromatogram. The analysis of a thin-layer chromatogram consists of determining the ratio of the distance each compound has traveled on the plate relative to the distance the solvent has traveled，the ratio of distances is called the R_f(retention factor). The distances must be measured from the same starting point，which is a physical characteristic of the compound. The R_f value for a compound depends on its structure as well as the adsorbent and mobile phase used (see Figure 2.9). Whenever a chromatogram is done，the R_f value should be calculated for each substance and the experimental conditions should be recorded in the laboratory notebook.

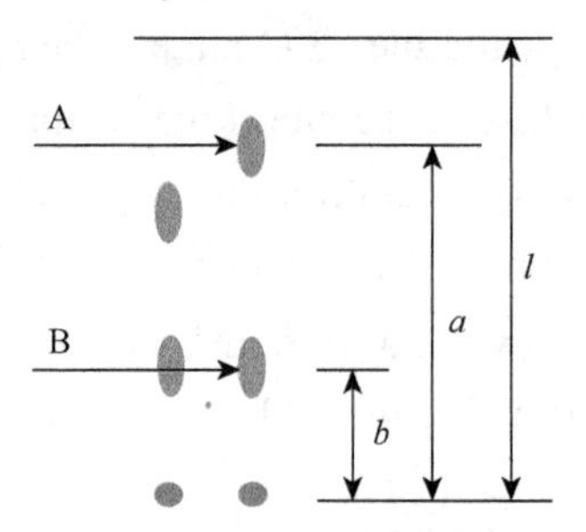

Figure 2.9　Determination of R_f value

2.8.2　Column chromatography

Column chromatography，which belongs to liquid chromatography，and the related methods of high performance liquid chromatography (HPLC) are part of the chromatographic methods so important in experimental organic chemistry. The principle of column chromatography is same as the TLC.

Column chromatography is generally used to separate compounds of low volatility，whereas gas chromatography(GC)works only for volatile mixtures. Unlike TLC and GC，liquid chromatography can be carried out with a wide range of sample quantities，ranging from a few micrograms for HPLC up to 10 g or more for column chromatography.

1. Elution solvents

Solvent choice is crucial for good separation. If the solvent to be used is strong polar，elution will be fast and there will be little separation between the product and impurities or between components of the mixture. If instead the solvent is weak polar or nonpolar，the compounds will be retained in the column. Therefore，several TLC plates should be developed with silica gel to determine the solvent (or solvent mixture) most suitable to keep the observed spots as far apart as possible.

2. Stationary phase

The amount of adsorbent (silica gel) employed depends on the amount of sample to be purified and on the R_f value of the spots on TLC. As a guide，the amount of silica gel per unit mass of sample would be 50：1. If too little adsorbent is used, the column will be overloaded and the separation will be poor. If too much adsorbent is used, the chromatography will take longer, require more elution solvent, and be no more efficient.

3. Chromatographic column

A height of 10-20 cm of silica gel often works well, and an 8：1 or 10：1 ratio of the adsorbent height to the inside column diameter is normal. Thus, a 1.5-2.5 cm column diameter is common for liquid chromatography on silica gel. The column is fixed with a clamp and a clamp holder to a lab stand, and the stationary phase is introduced by forming slurry with the same eluent to be employed after a powder funnel is used.

4. Sample application

Usually the sample is placed in the column, adjusting the eluent level just to the silica gel limit. If the sample is soluble in the eluent, it is dissolved in the minimum amount and transferred to the column with a dropper, on silica gel, avoiding splashing onto the walls of the column. In this case that the solvent is insoluble, in a round flask a solution of the sample to which silica gel is added is prepared. After the solvent evaporates, the silica gel with the product adsorbed on the surface thereof is introduced into the column.

5. Eluent techniques

Once the sample has been placed in the holder, the eluent is added and the eluent flow rate should be controlled in order to achieve an efficient separation. For each chromatographic separation, there is an optimum value of the elution rate. If the rate at which the eluent flows is too slow, the sample will diffuse excessively into the column, the run time will lengthen the separation time, and the efficiency of the separation will decrease. If instead the eluent flow is too fast, not enough sample will be provided to interact with the column packing, and the separation time will also be inefficient (see Figure 2.10).

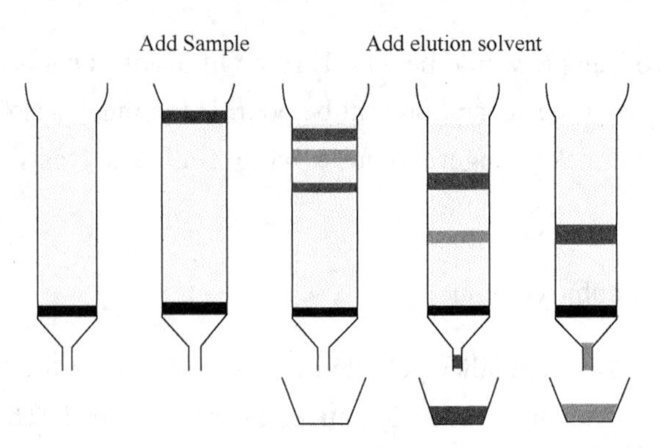

Figure 2.10 Column chromatography apparatus

6. Collecting eluent and isolating the products

The usual practice is to collect the eluent in a test tube. To find out which are the product fractions or products that have been purified，an analysis of these by TLC is made. Once the product is identified，the corresponding fractions are collected in a round flask and the solvent is removed（rotary evaporator）.

2.8.3 Paper chromatography

Paper chromatography（PC）is a partition chromatography. The filter paper is the carrier，the sample solutions are separated in the filter paper. Paper chromatography，the water or organic solvent adsorbed on the filter paper is the stationary phase. The containing water or organic solvents are mobile phase，also can be called the developing agent. It is a partition process between two phases. It can be used for the mixtures. The separation can be due to the different distribution coefficient of the components in stationary phase and mobile phase.

The operation of the process of paper chromatography and thin-layer chromatography is similar. Hold the micropipette vertically to the paper and apply the sample by touching the micropipette gently and briefly to the paper about 2-3 cm far from the bottom edge. Mark the edge of the paper with a pencil at the same height as the center of the spot，this mark indicates the starting point of the compound. A small amount of the mixture being separated is spotted on the filter paper. Then the filter paper is placed in a closed chamber（see Figure 2.11）. When the spots of the paper is dry，put the low edge of the filter paper into the developing agent. The solvent rises through the paper by capillary action. Finally the filter paper is picked up from the developing chamber when the solvent front is near the top of the paper. The position of the solvent front is marked immediately with a pencil，and then the solvent is evaporated.

The paper chromatography separations of colored compounds can be seen directly. However, many organic compounds are colorless, so an indirect visualization technique is needed. The paper can be illuminated by exposure to ultra violet radiation when it is impregnated with a fluorescent indicator or visualized reagent.

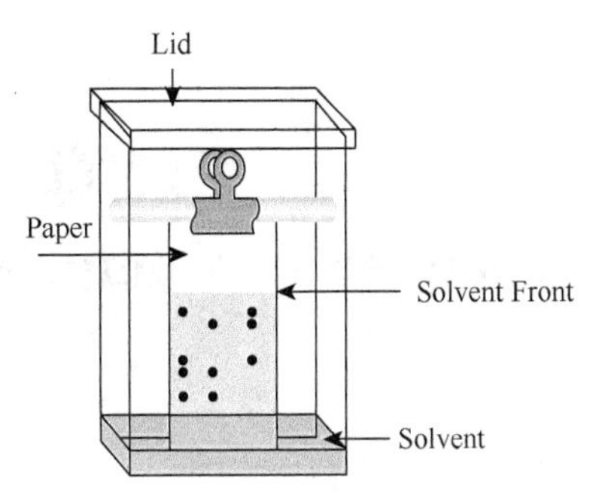

Figure 2.11 PC apparatus

Calculating R_f value of compounds, R_f value is related to the structure of separated compounds, the nature of stationary phase and mobile phase, temperature and quality of filter paper and other factors. Experimental data are often different from the literature due to the effecting factors of R_f value. So you must compare with standard substance.

The advantages of paper chromatography are very easy to operate and cheap: chromatogram can be stored for a long time, its disadvantage is slow to develop.

第3章　有机化学综合实验
Chapter 3　Comprehensive Experiments of Organic Chemistry

第一部分　天然产物提取与分离
Part 1　Extraction and Separation of Natural Products

3.1　咖啡因提取与分离

【实验目的】

（1）学习提取咖啡因的原理和方法。

（2）掌握提取和升华操作。

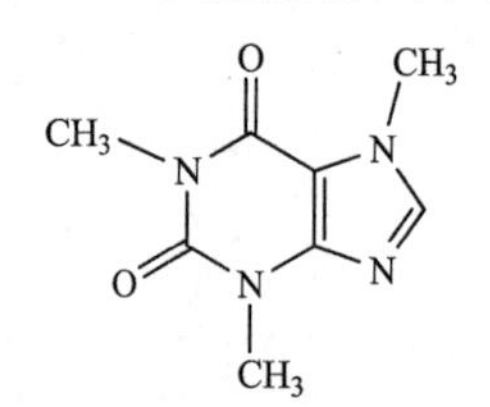

图3.1　咖啡因分子结构式

【实验原理】

咖啡因是存在于可可、茶叶、咖啡豆等植物中的一种苦味生物碱，为杂环化合物嘌呤的衍生物，化学名称是1, 3, 7-三甲基-2, 6-二氧嘌呤，其结构式如图3.1所示。

咖啡因是弱碱性化合物，常以盐或游离态存在，易溶于热水、$CHCl_3$、C_2H_5OH等。含结晶水的咖啡因是无色针状晶体，味苦，在100℃时即失去结晶水并开始升华，120℃时显著升华，至178℃时迅速升华。无水咖啡因的熔点为238℃。

茶叶中咖啡因的含量为1%～5%。本实验用水作溶剂从茶叶中提取咖啡因，然后浓缩水提液、焙炒得粗咖啡因，再通过升华法提纯。

【实验仪器与材料】

仪器：烧杯，玻璃漏斗，蒸发皿，脱脂棉，滤纸，量筒，电热板，减压过滤装置。

材料：茶叶，细砂，氧化钙。

【实验步骤】

1. 提取

称取 20 g 茶叶，加到盛有 200 mL 蒸馏水的烧杯中，电热板用大火加热，煮沸 30 min，其间需不断补加蒸馏水，以弥补蒸发掉的水分。然后趁热过滤，滤液置于烧杯中加入磁力搅拌子，在电加热磁力搅拌装置上加热、搅拌、浓缩至约 20 mL。

2. 升华

待滤液浓缩至 5～10 mL 时，液体转移至蒸发皿中，加入 4 g 氧化钙，搅拌，在砂浴上继续加热浓缩。待浓缩液呈黏稠状时，在蒸发皿上放一层有许多小孔的滤纸，上部再放一个干燥的玻璃漏斗，漏斗颈处塞有棉花，加热升华。当滤纸上出现针状结晶时，停止加热，冷却，观察晶体状态，称量，计算产率。

3.1　Extraction and Separation of Caffeine

【**Objectives**】

（1）To learn the theory and methods about extraction and separation of caffeine.

（2）To master the techniques of extraction and sublimation.

【**Principles**】

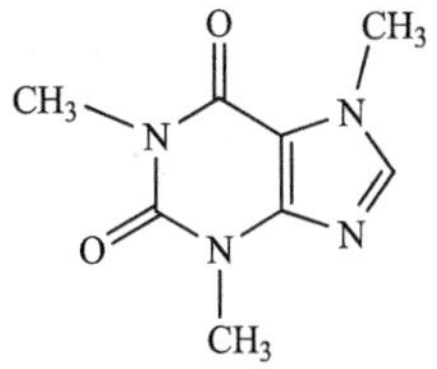

Caffeine is one of the main substances in coffee，tea and cocoa beans. The caffeine is a bitter alkaloid and derivative of purine，as a heterocyclic compound. The chemical name of caffeine is 1, 3, 7-trimethyl-1H-purine-2, 6(3H, 7H)-dione and the chemical structure is shown in Figure 3.1.

Figure 3.1　Caffeine

Caffeine is a weak alkaline compound，often in salt or free form，soluble in hot water，$CHCl_3$，C_2H_5OH，etc. Caffeine containing crystalline water is a colorless acicular crystal with bitter taste. It loses crystalline water at 100℃ and begins to sublimate. It distinctly sublimates at 120℃ and rapidly sublimates at 178℃. The melting point of anhydrous caffeine is 238℃.

As much as 1%-5% by weight of the leaf material in tea plants consists of caffeine. In this experiment，caffeine is extracted from tea with water as solvent，then concentrated from water extract，roasted to obtain crude caffeine，and then purified by sublimation.

【Apparatus and Materials】

Apparatus：beaker，glass funnel，evaporating dish，cotton，filter paper，measuring cylinder，electric heating plate，vacuum filtration apparatus.

Materials：tea，fine sand，calcium oxide.

【Procedure】

1. Extraction

Weigh 20 g of dry tea，then pack them into a beaker and place them on the electric heating plate. Add 200 mL water. Control the temperature at about 100℃ for 30 min. Then filter the solution by a vacuum filtration while hot. The filtrate is concentrated to 20 mL on the electric heating plate under magnetic stirring.

2. Sublimation

When the filtrate is concentrated to 5-10 mL，the liquid is transferred to the evaporating dish，adds 4 g of calcium oxide，stirs and continues to heat and concentrate in the sand bath. When the concentrated liquid is viscous，a layer of filter paper with many small holes is placed on the evaporating dish，and a dry glass funnel is placed on the top. Cotton is stuffed at the neck of the funnel. Continue to heat and sublimate. When acicular crystals appear on the filter paper，stop to heat and cool. Observe the crystal state，weigh and calculate the yield.

3.2　果皮中有效成分提取与分离

【实验目的】

（1）了解天然产物果胶的特点。

（2）掌握用酸提法从天然植物中提取果胶的基本原理和操作方法。

【实验原理】

果胶是一种天然有效成分，是人体所需七大营养素中膳食纤维的主要成分之一，主要存在于新鲜果皮（如香蕉、苹果、橘子等）中。它们常用来作为胶凝剂和增稠剂制作果酱、果冻、软糖、饮料、乳制品等。

果胶又称果胶多糖，富含半乳糖醛酸。果胶主要以不溶于水的原果胶形式存在于植物中，是一种线形的多糖聚合物，含有数百至约 1000 个脱水半乳糖醛酸残基，其主要成分是部分甲酯化的 *α*-1, 4-D-聚半乳糖醛酸，结构片段见图 3.2。

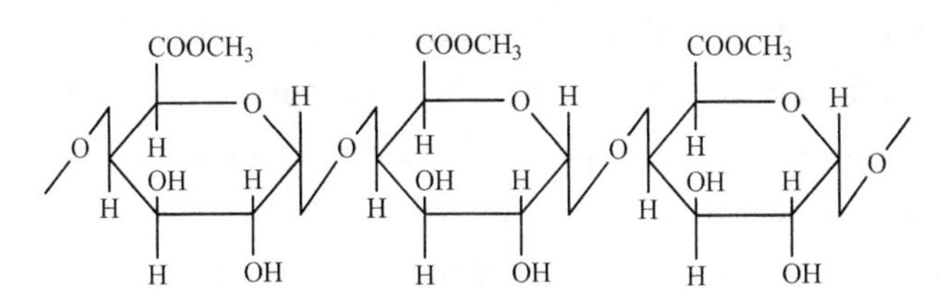

图 3.2　果胶分子结构片段

新鲜植物中果胶的典型含量为：苹果 1%～1.5%，杏 1%，樱桃 0.4%，橙子 0.5%～3.5%，胡萝卜 1.4%，柑橘皮 30%。果胶生产的主要原料是橘子皮或苹果渣，这两种都是果汁生产的副产品，甜菜渣也有少量使用。

当用热稀酸（pH 为 1.5～3.5）从植物中提取果胶时，不溶于水的原果胶被水解成果胶进入溶液。过滤后，萃取液浓缩，加入乙醇或异丙醇使果胶沉淀，然后将醇沉淀的果胶分离、洗涤和干燥。

【实验仪器与材料】

仪器：烧杯，量筒，滤纸，电热板，减压过滤装置。

材料：果皮（如香蕉、苹果、橘子等），浓盐酸，活性炭，95%乙醇。

【实验步骤】

取 10 g 新鲜果皮（如香蕉、苹果、橘子等）放入烧杯中，加 60 mL 水，再加入 1.5～2 mL 浓盐酸，电热板上加热至沸，在搅拌下维持沸腾约 30 min，用减压过滤法过滤除去残渣，滤液内加入少量活性炭，再加热 15 min，溶液变稠，再次减压过滤，得浅黄色滤液。

将滤液转移到 50 mL 小烧杯中，在不断搅拌下慢慢加入等体积的 95%乙醇，会看到出现絮状果胶沉淀，静置 3～5 min，减压过滤，并用 95%乙醇 2～3 mL 分 2～3 次洗涤沉淀，然后得果胶固体。

3.2　Extraction and Separation of Effective Components from Pericarp

【Objectives】

（1）To know the characteristic of the natural pectin.

（2）To master the principle and technique of extraction with the acid from a natural plant.

【Principles】

Pectin is a kind of natural effective component and one of the main components of dietary fiber in the seven nutrient elements of human body，which exists in the fresh peels（such as banana，apple，orange，etc.）. They，as gelling agent and thickening agent，were used to prepare jams，jellies，soft sweets，drinks，dairy products and so on.

Pectins，also known as pectic polysaccharides，are rich in galacturonic acid. Several distinct polysaccharides have been identified and characterized within the pectic group. Homogalacturonans are linear chains of *α*-(1-4)-linked D-galacturonic acid. Pectin structure fragment is shown in Figure 3.2.

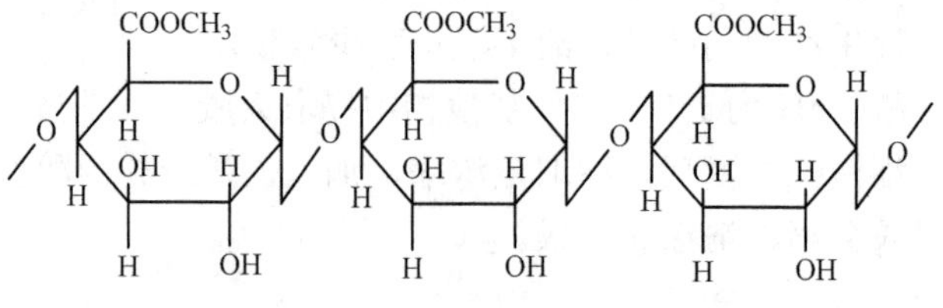

Figure 3.2　Pectin

Typical levels of pectin in fresh plants are：apple 1%-1.5%，apricot 1%，cherry 0.4%，orange 0.5%-3.5%，carota 1.4%，citrus peel 30%. The main raw materials for pectin production are dried citrus peels or apple pomace，both by-products of juice production. Pomace from sugar beets is also used to a small extent.

From these materials，pectin is extracted by adding hot dilute acid at pH-values from 1.5-3.5. During several hours of extraction，the protopectin loses some of its branching and chain length and goes into solution. After filtering, the extract is concentrated and the pectin is then precipitated by adding ethanol or isopropanol. Then alcohol-precipitated pectin is separated, washed and dried.

【Apparatus and Materials】

Apparatus: beaker, measuring cylinder, filter paper, electric heating plate, vacuum filtration apparatus.

Materials: fresh pericarp (such as banana, apple, orange, etc.), concentrated hydrochloric acid, acticarbon, 95% ethanol.

【Procedure】

Weigh 10 g of the fresh pericarp (such as banana, apple, orange, etc.), then pack them into a beaker, add 60 mL water, 1.5-2 mL concentrated hydrochloric acid and place them on the electric heating plate. Control the temperature at about 100℃ for 30 min. Then filter the solution by a vacuum filtration while it is hot. Heat the solution with a small amount of the acticarbon for 15 min until the solution turns thick, then filter the solution by a vacuum filtration again, you can get a little of yellow solution.

Transfer the solution to a 50 mL beaker. The solvent is mixed with an equal volume of 95% ethanol gently and stir rapidly. Let stand for 3-5 min, you can see the flocculation. Wash the pectin products by 2-3 mL ethanol (may occur 2-3 times) after a vacuum filtration. Collect the well-formed pectin products.

3.3　香精油的提取与分离

【实验目的】

（1）学习香料提取和分离的原理和方法。

（2）掌握旋转蒸发仪的操作方法。

【实验原理】

精油也称为挥发油、醚油，是一种浓缩疏水性液体，含有植物挥发性芳香化合物，如丁香油、肉桂油。精油通常存在于植物的花或果实中，但也可能存在于植物其他部位，如茎、叶或根中。植物精油可以通过挤压法、研磨法、溶剂法或蒸馏法等工艺提取。与脂肪油相比，精油涂到滤纸上时会完全蒸发，不会留下污渍（残留物）。

精油经常被用于香水、化妆品、肥皂和其他家用清洁产品中，或用于食品和饮料的调味。另一方面，使用香料和香精进行治疗或用于装饰也已有数千年的历史。芳香疗法是一种替代药物疗法，其疗效可归因于芳香化合物可能有助于引起放松，但没有足够的证据表明精油可以有效地治疗任何疾病。

旋转蒸发仪是化学实验室常用设备，主要用于减压条件下连续蒸馏易挥发性溶剂。通过程序控制，使蒸馏瓶在最适合速度下恒速旋转，以增大蒸发面积。通过真空泵使蒸馏瓶内处于负压状态，旋转蒸发器系统可以密封减压。蒸馏瓶在旋转的同时置于水浴锅中恒温加热，加热温度可接近该溶剂的沸点。瓶内溶液在负压、旋转下在蒸馏瓶内形成薄膜，增大蒸发面积，进行加热扩散蒸发。蒸发速率可以通过调节真空度、水浴温度和烧瓶的旋转速率来控制。

【实验仪器与材料】

仪器：圆底烧瓶，锥形瓶，烧杯，漏斗，量筒，研钵，电磁搅拌加热套，磁力搅拌子，回流装置，旋转蒸发仪，减压过滤装置。

材料：干燥的香料（肉桂或丁香），乙酸乙酯，无水硫酸钠。

【实验步骤】

1. 提取

称取 2.5 g 香料（肉桂或丁香），用研钵研碎，将固体转移到 50 mL 圆底烧瓶中，加入 20 mL 乙酸乙酯，加入磁力搅拌子，组装成回流装置，混合物回流 0.5 h。

将圆底烧瓶冷却至室温，滤液减压过滤到干燥锥形瓶中。用 5 mL 乙酸乙酯洗涤圆底烧瓶，以确保从香料中所有提取物过滤完全。

在锥形瓶中加入无水硫酸钠干燥溶液，然后将滤液过滤到干燥并已称量的 50 mL 圆底烧瓶中。

2. 分离

用旋转蒸发器在减压下除去溶剂，称量圆底烧瓶总质量，计算并确定起始样品中精油的含量。

3.3　Extraction and Separation of Essential Oils

【Objectives】

（1）To learn the theory and methods about extraction and separation for spices.

（2）To master the experimental technique of rotary evaporator.

【Principles】

The essential oil，which also known as a volatile oil，an ethereal oil is a concentrated hydrophobic liquid containing volatile aroma compounds from plant，such as oil of clove，oil of cinnamon. Essential oils are generally located in the seeds or flowers but may exist in other plant parts such as stems，leaves，or roots. They can be isolated from a plant by various processes or a combination of processes，such as mechanical pressing，grinding，solvent extraction，or distillation. In contrast to fatty oils，essential oils evaporate completely without leaving a stain（residue）under the filter paper.

The essential oils are used in perfumes，cosmetics，soaps and other household cleaning products，or in flavoring food and drink. On the other hand，the use of fragrances and essences for healing or for ornamental purposes has also been known for thousands of years. Aromatherapy，a form of alternative medicine，whose healing effects are ascribed to aromatic compounds，may be useful to induce relaxation，but there is not sufficient evidence that essential oils can effectively treat any condition.

In chemical laboratory，solvents are removed under reduced pressure，using an apparatus called a rotary evaporator. This device is especially designed for the rapid evaporation of solvents without bumping. A variable-speed motor is used to rotate the flask containing the solvent being evaporated. While the flask is being rotated， a vacuum is applied and the flask is heated. The rotation of the flask spreads a thin film of the solution on the inner surface of the flask to accelerate evaporation，and it also agitates the contents of the flask to reduce the problem of bumping. The rate of evaporation can be controlled by adjusting the vacuum，the temperature of the water bath，and the rate of rotation of the flask.

【Apparatus and Materials】

Apparatus：round-bottom flask，Erlenmeyer flask，beaker，funnel，measuring cylinder，mortar，magnetic stirring electric heating mantle，stir bar，reflux apparatus，rotary evaporator，vacuum filtration apparatus.

Materials：dried spice（cinnamon or clove），ethyl acetate，anhydrous sodium sulfate.

【Procedure】

1. Extracting

Weigh 2.5 g of spice（cinnamon or clove），and finely grind using a mortar. Transfer the solid to a 50 mL round-bottom flask，add 20 mL of ethyl acetate，and add a stir bar. Fit the reflux apparatus，reflux the mixture for 0.5 h.

Cool the flask contents to room temperature and filter by vacuum filtration apparatus into a dry Erlenmeyer flask. Wash the round-bottom flask with 5 mL of ethyl acetate，and pour the solvent onto the solid remaining in the filter paper to ensure that all the extracts from the spice are obtained.

Then dry the ethyl acetate solution over anhydrous sodium sulfate，and filter by gravity into a 50 mL tared and dried round flask.

2. Separation

Remove the solvent under reduced pressure（rotary evaporator），weigh the contents，and determine the percentage of the essential oil from the starting sample.

3.4　纸色谱分离、鉴定氨基酸

【实验目的】

（1）学习纸色谱的基本原理。

（2）掌握分离氨基酸的方法。

【实验原理】

将含有氨基酸的样品点到滤纸上，再将滤纸放到装有少量展开剂的层析缸中。混合物中不同的氨基酸随着溶剂在滤纸上展开，在滤纸上显出不同的斑点，每个斑点表示不同的化合物，这就是纸色谱。吸附在滤纸上的水被称为固定相，溶剂称为流动相。纸色谱可用来分离混合物，利用物质在固定相和流动相中的分配系数的不同而进行分离。

纸色谱法主要用于分离有色物质，目前也可用于分离无色物质如氨基酸。茚三酮与氨基酸反应显紫色。所以，纸色谱展开后，喷上茚三酮可显出不同的氨基酸，然后测定 R_f 值。

【实验仪器与材料】

仪器：层析缸，滤纸，毛细管，镊子，烘箱，喷壶，格尺，铅笔。

材料：谷氨酸，亮氨酸，茚三酮，正丁醇，乙酸。

【实验步骤】

选滤纸一条（5 cm×20 cm）（只能接触一端），置于一洁净的纸面上，用铅笔在一端从左到右轻轻画一条线（不能用钢笔!），平行从底边 1.5 cm 处画三点，两点之间距离 1.5 cm，表示 A、A + B 及 B。分别用毛细管点取谷氨酸、亮氨酸样品，点在标点 A、B 上，混合样在 A + B 上。滤纸干燥几分钟。如果样品点量不足，可在同一位置重复操作。

按比例向层析缸中倒入展开液，溶液高度约 1.5 cm，为了形成饱和蒸气，在展开前要将层析缸密闭几分钟。用铜丝把滤纸条挂在缸中，点样线在下端（点样线不能浸在展开剂里，滤纸边不能接触缸壁），然后轻轻盖上层析缸，进行展开。

纸色谱展开，当展开完成时，应迅速取出滤纸，在溶剂干燥前标出溶剂前沿的位置。然后用茚三酮显色，在 105℃烘箱干燥，

取出纸色谱滤纸条，放到桌面上，标出每个斑点，测出原点到层析点中心的距离及原点到溶剂前沿的距离。计算 R_f 值，分析确定结果。

3.4 Separation and Identification of Amino Acids by Paper Chromatography

【Objectives】

（1）To learn the principles of paper chromatography.

（2）To master the method of separating amino acid.

【Principles】

Paper chromatography is used when a small spot containing a mixture of amino acid is placed near the bottom of a piece of filter paper，and the filter paper is placed in a covered jar containing a small amount of suitable solvent；the solvent moves up the filter paper，carrying each amino acid on the paper to a different extent. This results in a series of spots on the paper，each spot corresponding to a different compound. The water absorbed on the filter paper is called the stationary phase，and the solvent is called the mobile phase. Paper chromatography can be used for the separation of mixture. The separation can be due to the different distribution coefficients of the components in stationary phase and mobile phase. Paper chromatography is primarily application.

Chromatography is primarily applied to separate colored substances，and now it may also be applied to separate colorless compounds，such as amino acid. The compound ninhydrin can react with all amino acids to produce purple products. So after developing the chromatogram，ninhydrin solution should be sprayed so that the spots，corresponding to the different amino acids，can be showed，you will then determine the R_f value for each compound.

【Apparatus and Materials】

Apparatus：chromatographic tank，filter paper，capillary tube，tweezers，oven，spray box，ruler，pencil.

Materials：glutamic acid，leucine，ninhydrin，*n*-butyl alcohol，acetic acid.

【Procedure】

Choose a precut（5 cm×20 cm）sheet（make sure to touch it only along the top edge）and place it on a clean sheet of paper. With the short way，using a straightedge，draw a light pencil line（Not ink !）from left to right. Parallel to and up from the bottom edge by 1.5 cm, place 3 small pencil marks at 1.5 cm intervals along this line. Label the three marks with A， A + B and B. Then clip the capillary tubes into the glutamic acid solution and make the spot on the sheet， applying it at the position marked with A, leucine at the spot B and the mixture of the two at the spot A + B. Allow the paper to dry for a few minutes. If the spot is not big enough， make second application at the same position as the first. Allow the paper to dry.

Pour the developers into the chromatographic tank at a depth of 1.5 cm. Allow it to stand for a few minutes before the separation, in order to form the saturated vapor in the jar. Hang the strip in the jar with the marked end at the lower end, and dip into the developer（insuring the spot do not dip into the developer and the edges of the paper are not allowed to touch the inside wall of the chromatography jar）. Cover the jar tightly.

Allow the chromatogram to develop. When the development is finished， you should remove the chromatogram and immediately mark the location of the solvent before it has a chance to dry. Dry it and spray it with ninhydrin solution. Then put your paper in an oven with a temperature of 105℃.

Remove the chromatogram， and take it back to your desk. Circle each spot and measure the distance from the origin to the center of each spot as well as the distance from the origin to the solvent front. Calculate the R_f values and determine each compound by R_f value.

3.5　纸电泳分离、鉴定氨基酸

【实验目的】

（1）学会氨基酸电泳的基本原理。

（2）练习用电泳法分离、鉴定氨基酸。

【实验原理】

带正电荷或负电荷的离子在电场中可以移向与其电性相反的电极，称为电泳。以滤纸为支撑物的电泳称为纸电泳。

氨基酸的结构取决于溶液的pH，氨基酸以两性离子存在时的pH称为等电点（pI）。在等电点，氨基酸的电荷为零，是电中性的，在电场中不移动；当溶液的pH大于等电点时，氨基酸将带负电荷，移向正极；反之，当溶液的pH小于等电点时，氨基酸将带正电荷，移向负极。因此，按不同氨基酸离子移动方向的不同，可以分离、鉴定氨基酸。

【实验仪器与材料】

仪器：DY-2电泳仪，滤纸，毛细管，烘箱，喷壶，镊子，格尺，铅笔。

材料：谷氨酸，精氨酸，茚三酮，巴比妥缓冲溶液（pH = 8.9）。

【实验步骤】

将滤纸条（5 cm×30 cm）放在一张洁净的纸上，用铅笔在滤纸中间画一条线（不能用钢笔！），标记三个点A、B和C，间距为1.5 cm。注意：不要用手指接触滤纸！用毛细管点样，使其干燥几分钟。再复点1～2次，再使其干燥。样品A是谷氨酸，B是精氨酸，C是混合样。

将点好的滤纸放在电泳仪的架子上，滤纸的两端滴上缓冲液，用缓冲液润湿滤纸点样线1 cm外的部分直到溶液渗透点样线，然后盖上电泳仪。

打开电源，调节电压在220～280 V范围内，约30 min后，关闭电源。

电泳完成后，用镊子取出滤纸挂起，在105℃烘干，然后用茚三酮显色，在105℃再干燥直至呈现斑点，圈出斑点。计算R_f值，分析确定结果。

3.5　Separation and Identification of Amino Acids by Paper Electrophoresis

【Objectives】

（1）To learn the principle of amino acids by paper electrophoresis.

（2）To practise the separation and identification of amino acid by paper electrophoresis.

【Principles】

Ions，which have positive or negative charges，will move towards the electrodes which have opposite charges. This quality is called electrophoresis. When used filter paper as a supporter in electrophoresis is paper electrophoresis（PE）.

As we all know，the exact structure of an amino acid depends on the pH. The pH at which an amino acid exists as zwitterion is known as the isoelectric point（pI）. At the pI，amino acids have a zero net charge and are electrically neutral and without removal in the electric field. When the pH of the solution is bigger than the pI of the compound，the amino acid will carry a net charge（anionic form），and will remove to the anode. Conversely，when pH of the solution is below the pI value，the amino acid exists predominantly in the cationic form（not positive charge），and to the cathode. According to the difference of the removal direction and the rate of different amino acid ions，we can separate and identify the amino acid.

【Apparatus and Materials】

Apparatus：DY-2 electrophoresis apparatus，filter paper，capillary tube，oven，spray box，ruler，pencil.

Materials：glutamic acid，arginine，ninhydrin，barbital buffer solution（pH = 8.9）.

【Procedure】

A precut sheet of filter paper（5 cm×30 cm）is placed on a clean sheet of paper. You need to draw a thin pencil（not ink!）line in the middle of the filter paper. Place three small pencil marks，label the marks whit A，B and C at the same intervals（1.5 cm）along this line，be careful not to touch the paper with your fingers. Then dip the capillary tube into the amino acid solutions and make the spots on the large sheet，apply them at the positions marked earlier. Allow the paper to dry for a few minutes，and then make a second application at the same position as the first. Allow the paper to dry. Spot A is glutamic acid，B is arginine and C is the mixture of the two.

Place the paper on the shelf of electrophoresis apparatus，dip both the ends of the paper into the buffer. Then humidify the paper out of the pencil line about 1 cm with buffer until the solution permeates the line and then cover the cell.

Turn on the power，adjust the voltage between 220 V and 280 V. About 30 min later，turn off the power.

When electrophoresis is finished，remove the strip out of the electrophoresis tank with a pair of tongs，and hang it in an oven at 105℃ to dry them. Then spray them with ninhydrin solution，put them in the oven and keep it under 105℃ until the spots emerge，remove the strips and circle each spot. Calculate the R_f values and determine each compound by R_f value.

3.6 薄层色谱法分离、鉴定叶绿素

【实验目的】

（1）学习用薄层色谱法分离、鉴定叶绿素。

（2）掌握薄层色谱操作技术。

【实验原理】

薄层色谱法是以薄层板作为载体，使样品溶液在薄层板上展开而达到分离的目的的分析方法，也称薄层层析。它是一种简单、价廉、快速、灵敏和高效快速分离和定性分析少量物质的实验技术，使用广泛，可用于化合物鉴定、跟踪反应进程等。

薄层色谱法与柱色谱法原理相同，且都是液-固吸附色谱。在薄层色谱中，固定相以薄层（约 250 μm）形式涂在玻璃板或其他硬质薄板上。分离的物质被点在薄板上，然后把薄板放入盛有足够展开剂的层析缸中（不能超过点样线），混合物中组分则以不同的速率随着溶剂展开、分离，通过计算各成分的 R_f 值加以鉴定。

菠菜叶含有多种天然产物，如胡萝卜素、叶绿素、叶黄素、可溶性维生素等。菠菜叶含有的 β-胡萝卜素和叶绿素，还有少量的叶黄素成分是菠菜叶显现颜色的主要原因。

本试验采用薄层色谱法对菠菜叶色素进行提取、分离、鉴定。

【实验仪器与材料】

仪器：研钵，分液漏斗，层析缸，毛细管，镊子，格尺，铅笔，锥形瓶，旋转蒸发仪。

材料：石油醚（b.p. 30～60℃），乙醚，新鲜菠菜，无水硫酸钠，氯仿，薄层板。

【实验步骤】

在研钵中，放入 10～15 g 新鲜菠菜叶和 10 mL 体积比为 2∶1 的石油醚和乙醚混合溶液，仔细研磨。用吸管将液体转入分液漏斗，加等体积水，旋摇（注意：如振摇液体可能乳化！）。静置分层，弃去下面的水层。有机层再用水洗涤两次，水相弃掉。将有机层倒入锥形瓶，加 2 g 无水硫酸钠干燥。放置几分钟后倾出溶液，将溶液浓缩。然后取 10 cm×2 cm 薄层板，用毛细管将带色溶液点在硅胶板上，样品点距末端约 1.5 cm。点样时，应避免样品点直径扩散超过 2 mm。将硅胶板晾干，用氯仿作展开剂，放入层析缸中展开。

展开后的薄层板上有时可能会有八个有色斑点，按 R_f 值递减，这些斑点已被

确证为：胡萝卜素（两个点，橙色）、叶绿素 a（蓝绿）、叶黄素（四个点，黄色）和叶绿素 b（绿色）。

计算各斑点的 R_f值，对应确认各物质。

3.6 Separation and Identification of Chlorophyll by Thin-Layer Chromatography

【Objectives】

（1）To learn the separation and identification of green leaf pigments by thin-layer chromatography（TLC）.

（2）To master the operation technique of TLC.

【Principles】

TLC is one of the most widely used analytical techniques. TLC is a simple，inexpensive，fast，sensitive，and efficient method for determining the number of components in a mixture，for possibly establishing whether or not two compounds are identical，and for following the process of reaction.

TLC involves the same principles as column chromatography，and it also is a form of liquid-solid adsorption chromatography. In this case，however，the solid adsorbent is spread as a thin layer（approximately 250 μm）on a plate of glass or rigid plastic. A drop of the solution to be separated is placed near one edge of the plate，and the plate is placed in a container，called a developing chamber，with enough of the eluting solvent to come to a level just below the "spot". The solvent migrates up the plate，carrying with it the components of the mixture at different rates. The result may then be seen as a series of spots on the plate，falling on a line perpendicular to the solvent level in the container. The retention factor（R_f）of a component can then be measured as indicated in the thin layer.

Spinach leaves contain a number of natural products such as carotene，chlorophyll，xanthophyll，soluble vitamins，etc. Spinach leaves present *β*-carotene and chlorophyll，these beings are primarily responsible for the leaf color，together with minor amounts of xanthophyll components.

In this experiment，the pigments from spinach leaves are extracted and separated by means of TLC.

【Apparatus and Materials】

Apparatus：mortar，separating funnel，chromatographic tank，capillary tube，tweezers，ruler，pencil，Erlenmeyer flask，rotatory evaporator.

Materials：petroleum ether（b.p. 30-60℃），diethyl ether，spinach，anhydrous sodium sulfate，chloroform，thin layer plate.

【Procedure】

Place in 10-15 g fresh spinach leaves and 10 mL of a 2∶1 mixture of petroleum ether and ethanol，and grind the leaves well with a pestle. By means of a pipet，transfer the liquid extract to a small separatory funnel and swirl with an equal volume of water（Note：shaking may cause formation of an emulsion!）. Separate and discard the lower aqueous. Repeat the water washing twice，discarding the aqueous phase each time. Transfer the petroleum ether layer to a small flask and add 2 g of anhydrous sodium sulfate. After a few minutes，separate solution from the sodium sulfate and concentrate it. Take a 10 cm×2 cm strip of silica gel chromatogram sheet. Place a spot of the pigment solution on the sheet about 1.5 cm from one end，using a capillary tube to apply the spot. Avoid allowing the spot to diffuse to a diameter of no more than 2 mm during the application of the sample. Allow the spot to dry，chloroform was used as the developer and develop in the chromatographic cylinder.

It is sometimes possible to observe as many as eight colored spots. In the order of decreasing values，these spots have been identified as the carotenes（two spots，orange），chlorophyll a（blue-green），the xanthophylls（four spots，yellow），and chlorophyll b（green）.

Calculate the R_f values of any spots observed on your developed plate and determine each compound by R_f value.

第二部分　有机化合物的制备
Part 2　Preparation for Organic Compounds

3.7　甲基橙的制备

【实验目的】

（1）学习甲基橙的合成原理和方法。

（2）掌握有机化学实验中反应温度的控制方法和熟悉减压过滤等基本操作。

【实验原理】

甲基橙是一种常用的酸碱指示剂，变色范围为3.1～4.4，颜色由红变黄，制备甲基橙最常用的方法是先将对氨基苯磺酸制成重氮盐，然后在低温、弱酸性条件下，与*N*, *N*-二甲基苯胺进行偶联，生成偶氮化合物甲基橙。这样合成得到的甲基橙是有杂质的粗品，可以通过重结晶等方法进行精制。

多步合成反应如下：

$$H_2N-C_6H_4-SO_3H + NaOH \longrightarrow H_2N-C_6H_4-SO_3Na + H_2O$$

$$H_2N-C_6H_4-SO_3Na \xrightarrow[HCl]{NaNO_2} \left[HO_3S-C_6H_4-\overset{+}{N}\equiv N\right]Cl^-$$

$$\xrightarrow[HAc]{C_6H_5N(CH_3)_2} \left[HO_3S-C_6H_4-N=N-C_6H_4-N(CH_3)_2\right]^+ Ac^-$$

$$\xrightarrow{NaOH} NaO_3S-C_6H_4-N=N-C_6H_4-N(CH_3)_2 + NaAc + H_2O$$

【实验仪器与材料】

仪器：烧杯，试管，量筒，温度计，滤纸，电磁搅拌加热板，减压过滤装置。

药品：对氨基苯磺酸，亚硝酸钠，*N*, *N*-二甲基苯胺，浓盐酸，氢氧化钠，冰醋酸，冰块，碘化钾-淀粉试纸。

【实验步骤】

1. 重氮盐溶液的制备

在 50 mL 烧杯中加入 6 mL 水，然后将 0.80 g 亚硝酸钠溶于水中，制成亚硝酸钠溶液备用。

在 100 mL 烧杯中加入 2.1 g 对氨基苯磺酸，用 10 mL 1.25 mol·L^{-1} 氢氧化钠溶液使其溶解，可适当加热，溶解后倒入已配好的亚硝酸钠溶液中，不断搅拌，并将烧杯放入冰-盐浴冷却到 0～5℃。然后在低温并不断搅拌下将 3 mL 浓盐酸和 10 mL 水配成的溶液缓缓滴加到上述混合液中，继续搅拌 10 min，此时溶液呈红橙色。用碘化钾-淀粉试纸检验重氮化反应的终点，若试纸出现蓝色，表示反应已达到终点。此时制得氨基苯磺酸的重氮盐溶液中往往有细小的白色晶体析出。把溶液保存在冰-盐浴中，待下步偶合反应中使用。

2. 偶合反应

在一支试管中加入 1.2 mL *N*, *N*-二甲基苯胺（避免接触皮肤！）和 1 mL 冰醋酸，摇匀后慢慢倾入上述已制备的重氮盐溶液中（在冰-盐浴中操作），边加边搅拌，加完后继续搅拌 10 min，然后加入约 30 mL 1.25 mol·L^{-1} 氢氧化钠溶液，用碘化钾-淀粉试纸检验，直到反应物变为橙色，减压过滤，得到碱式甲基橙粗产物。

3. 产物验证

取少量甲基橙产品在一小试管中用蒸馏水溶解。在另外 2 支试管中分别加入 3～5 mL 稀盐酸和稀氢氧化钠溶液，每支试管中滴加 1～3 滴甲基橙溶液，观察并解释溶液颜色变化。

3.7　Preparation for Methyl Orange

【Objectives】

（1）To learn the principle and method of preparing methyl orange.

（2）To master how to control the reaction temperature and be familiar with the operation of vacuum filtration.

【Principles】

Methyl orange is a pH indicator and it is very often used in titrations due to its clear color change along with the variation of pH value（3.1-4.4）. In the acid condition it is reddish and in the alkali condition it is yellow. In this experiment, methyl orange is synthesized by a diazonium coupling reaction of diazotized benzene sulfonic acid and *N*, *N*-dimethylaniline. By this method, the methyl orange synthesized is a crude product with impurities, which can be refined by recrystallization or other methods.

The main reactions:

$$H_2N-C_6H_4-SO_3H + NaOH \longrightarrow H_2N-C_6H_4-SO_3Na + H_2O$$

$$H_2N-C_6H_4-SO_3Na \xrightarrow[HCl]{NaNO_2} \left[HO_3S-C_6H_4-\overset{+}{N}\equiv N\right]Cl^-$$

$$\xrightarrow[HAc]{C_6H_5N(CH_3)_2} \left[HO_3S-C_6H_4-N=N-C_6H_4-N(CH_3)_2\right]^+ Ac^-$$

$$\xrightarrow{NaOH} NaO_3S-C_6H_4-N=N-C_6H_4-N(CH_3)_2 + NaAc + H_2O$$

【Apparatus and Materials】

Apparatus: beaker, test tube, measuring cylinder, thermometer, filter paper, magnetic stirring electric heating plate, vacuum filtration apparatus.

Materials: *p*-aminobenzene sulfonic acid, sodium nitrite, *N*, *N*-dimethylaniline, concentrated hydrochloric acid, sodium hydroxide, acetic acid, ice, potassium iodide-starch test paper.

【Procedure】

1. Diazotization reaction

In a 50 mL beaker, add 0.80 g of sodium nitrite into 6 mL of water to make the solution.

In a 100 mL beaker, add 2.1 g of *p*-aminobenzene sulfonic acid and 10 mL of 1.25 $mol \cdot L^{-1}$ sodium hydroxide aqueous solution. Heat the solution in a hot water bath until the solids completely dissolved. After dissolving, pour into the prepared sodium nitrite solution, stir and cool the beaker in an ice-salt bath to 0-5℃. Add a mixture of 3 mL concentrated HCl and 10 mL of water dropwise under swirling. Keep this solution cold（0-5℃）in the ice-salt bath all the time and continue to stir for about 10 min. The solution should turn red-orange at this stage. Potassium iodide-starch test paper is used to test the end point of diazotization reaction. If the test paper appears blue，the reaction has reached the end point. At this time，small white crystals often precipitate in the diazonium salt solution of aminobenzene sulfonic acid. The solution is stored in an ice-salt bath and will be used in the next coupling reaction.

2. Coupling reaction

Mix 1.2 mL of *N*, *N*-dimethylaniline（Absorbed through the skin! Handle only with gloves!）and 1 mL of acetic acid in a test tube，then slowly add，with constant stirring，this solution to the diazonium salt suspension in the beaker under an ice-salt bath. As the dull，red-purple solid starts to appear，stir continuously for 10 min to ensure a complete reaction. 30 mL of 1.25 $mol \cdot L^{-1}$ sodium hydroxide aqueous solution is slowly added until the solution turns orange，the potassium iodide-starch test paper is used to detect，and the crude alkaline methyl orange is obtained by vacuum filtration.

3. Product validation

Dissolve a small amount of methyl orange crystals in distilled water in a test tube. Take two test tubes. Add 3-5 mL of dilute HCl solution and dilute NaOH solution respectively，and then add about 1-3 drops of methyl orange solution in each tube by shaking. Observe and explain the color change of the solution.

3.8　乙酸异戊酯的制备

【实验目的】

（1）学习乙酸异戊酯的合成原理和方法。

（2）掌握回流、萃取基本操作。

【实验原理】

酯化反应是一类重要有机化学反应，是醇与羧酸或无机酸生成酯和水的反应，可分为三类：羧酸与醇，无机含氧酸与醇和无机强酸与醇反应。酯化反应可逆，且一般反应极缓慢，故常用浓硫酸作催化剂。典型的酯化反应有乙醇和乙酸反应，生成具有芳香气味的乙酸乙酯，乙酸乙酯是制造染料和医药的原料。

乙酸异戊酯是一种有机酸酯，因其具有香蕉的香味，故又称香蕉水，乙酸异戊酯可用乙酸和异戊醇直接酯化的方法合成。由于酯化反应可逆，为了使反应进行得比较完全，通常使其中某一种反应物过量，或者不断移去某一种生成物，促使反应向右移动，以利于产物的生成。由于乙酸比异戊醇便宜，而且易从反应混合物中除去，实验中通常使用过量乙酸进行制备。

主反应式如下：

$$CH_3COOH + (CH_3)_2CHCH_2CH_2OH \xrightleftharpoons{H_2SO_4} CH_3COOCH_2CH_2CH(CH_3)_2 + H_2O$$

【实验仪器与材料】

仪器：圆底烧瓶，水浴锅，分液漏斗，锥形瓶，量筒，烧杯，电磁搅拌加热套，回流装置。

材料：异戊醇，冰醋酸，浓硫酸，碳酸钠，氯化钠，无水硫酸镁。

【实验步骤】

向干燥的 100 mL 圆底烧瓶中加入 18 mL 异戊醇、24 mL 冰醋酸，然后缓慢分批地加入 5 滴浓硫酸，并加入磁力搅拌子，安装回流冷凝管，加热回流。

反应完毕后，取下圆底烧瓶，在冰水浴中冷却至 10℃，然后倒入 125 mL 分液漏斗中（注意：不能使磁力搅拌子进入分液漏斗！）。振荡几分钟，静置分层，弃去下层水溶液，若未分层可加入 20 mL 饱和食盐水使其分层后分离。再少量多次用 10%碳酸钠溶液进行洗涤（注意放气），至上层溶液不再呈酸性为止。用锥形瓶收集萃取后的产品，用无水硫酸镁干燥，称量，计算产率。

3.8 Preparation for Isoamyl Acetate

【Objectives】

（1）To learn the principle and method of preparing isoamyl acetate.

（2）To master the fundamental operations of reflux and extraction.

【Principles】

Esterification reaction，which is the reaction of alcohol with carboxylic acid or inorganic acid to form ester and water，is one type of the important organic chemical reaction. The esterification reaction can be divided into three categories：carboxylic acid and alcohol，inorganic oxyacid and alcohol，inorganic strong acid and alcohol. It is reversible and generally very slow，so concentrated sulfuric acid is often used as catalyst. The typical esterification reaction is the reaction of ethanol and acetic acid to produce ethyl acetate with aromatic odor，which is the raw material for manufacturing dyeing and medicine.

Isoamyl acetate is an organic ester. Because of its banana flavor it is also called banana water. Isoamyl acetate can be synthesized by direct esterification of acetic acid and isoamyl alcohol. Because the esterification reaction is reversible，in order to make the reaction complete，one of the reactants is usually excessive，or one of the products is constantly removed，so that the reaction moves to the right to facilitate the formation of products. Because acetic acid is cheaper than isoamyl alcohol and can be easily removed from the reaction mixture，excessive acetic acid is usually used in the experiment.

The main reaction：

$$CH_3COOH + (CH_3)_2CHCH_2CH_2OH \xrightleftharpoons{H_2SO_4} CH_3COOCH_2CH_2CH(CH_3)_2 + H_2O$$

【Apparatus and Materials】

Apparatus：round-bottom flask，water-bath，separating funnel，Erlenmeyer flask，measuring cylinder，beaker，magnetic stirring electric heating mantle，reflux apparatus.

Materials：isoamyl alcohol，glacial acetic acid，concentrated sulfuric acid，sodium carbonate，sodium chloride，anhydrous magnesium sulfate.

【Procedure】

Mix 18 mL of isoamyl alcohol and 24 mL of glacial acetic acid into a dry

100 mL round-bottom flask. Slowly add 5 drops of concentrated sulfuric acid in batches to the contents of the flask with swirling. Add a stir bar to the mixture. Fit a reflux condenser to the flask， and heat the mixture under reflux.

After the reaction is completed， remove the round-bottom flask， cool it in the ice water bath to 10℃，and then pour the product into the 125 mL separating funnel（Note： The stir bar can not enter the separating funnel!）. Shake it several times， keep static stratification， discard the lower water solution. If the solution is not stratified， 20 mL saturated salt water can be added after stratification. Wash the organic layer with 10% of the sodium carbonate solution several times in a small amount（Note：deflate!），until the upper solution is no longer acidic. The extracted product is collected in the Erlenmeyer flask and dried with anhydrous magnesium sulfate. Weigh and calculate the yield.

3.9 乙酰水杨酸的制备

【实验目的】

（1）学习酰化反应的原理和阿司匹林的制备方法。

（2）掌握有机合成中固体产物的分离、提纯方法，巩固重结晶和减压过滤基本操作。

（3）学习通过测定熔点和化学试剂检测产品纯度的方法。

【实验原理】

阿司匹林即乙酰水杨酸，是销售最广泛的非处方药。它能退烧（解热），减轻疼痛（止痛药），减轻肿胀、酸痛和发红（消炎药）。尽管有副作用，阿司匹林仍然是目前最安全、最便宜和最有效的非处方药。

本实验以浓硫酸为催化剂，水杨酸与乙酸酐反应制备阿司匹林。副反应主要是酚羟基和羧基之间的脱水反应，当温度高于 90℃时，分子间脱水加剧，从而生成不溶于水的聚合物。反应生成的粗制乙酰水杨酸含有未反应完的水杨酸以及一些相对分子质量较大的副产物，可采用醇-水混合溶剂进行重结晶提纯。或先在碱溶液中溶解酸，过滤除去不溶物后再酸化，使乙酰水杨酸重结晶析出。

主反应式如下：

$$\text{C}_6\text{H}_4(\text{COOH})(\text{OH}) + (\text{CH}_3\text{CO})_2\text{O} \xrightarrow{\text{H}_2\text{SO}_4} \text{C}_6\text{H}_4(\text{COOH})(\text{OCOCH}_3) + \text{CH}_3\text{COOH}$$

由于很少量的酚羟 基就能与 Fe^{3+}在水溶液中形成紫红色的配合物，所以用 $FeCl_3$ 溶液检验乙酰水杨酸是否提纯是实验室最简单、常用的方法之一，有时也通过测定熔点来检测产物纯度。

【实验仪器与材料】

仪器：锥形瓶，烧杯，水浴锅，量筒，试管，熔点仪，电磁搅拌加热板，减压过滤装置。

材料：水杨酸，乙酸酐，浓硫酸，浓盐酸，碳酸氢钠，三氯化铁。

【实验步骤】

1. 合成

水浴锅中装适量水，加热至 80～90℃，作为反应水浴装置。在 25 mL 干燥的锥形瓶中加入 2.0 g 水杨酸和 5 mL 乙酸酐，然后加入 4 滴浓硫酸，充分振摇，放置于水浴中加热溶解后继续振摇 8～10 min，然后将溶液倒入 100 mL 烧杯中，再加 30 mL 冰水冷却使结晶完全析出，减压过滤，用少量（约 5 mL）蒸馏水洗涤固体两次。

2. 纯化

将过滤得到的粗品置于 50 mL 烧杯中，加入 30 mL 饱和碳酸氢钠溶液，充分搅拌后减压过滤除去不溶物。滤液用浓盐酸酸化后，有晶体析出，用冰水冷却使结晶析出完全。结晶经减压过滤、水洗、红外干燥后，即得纯乙酰水杨酸，称量，计算产率。

3. 产品纯度的检测

（1）取 3 支试管并编号，1 号试管中放入少量水杨酸，2 号试管中放入少量重结晶纯化后的产品，3 号试管中放入少量乙酰水杨酸标准品。分别加入 5 mL 蒸馏水使样品完全溶解，然后分别滴入 2～3 滴 1%三氯化铁溶液，观察现象并记录。

（2）产品干燥后研细，用熔点仪测定样品熔点。纯净乙酰水杨酸固体的熔点为 135～136℃。

3.9 Preparation for Acetylsalicylic Acid

【Objectives】

（1）To learn the method of preparing aspirin by acetylation reaction.

（2）To master the techniques of separation and purification of solid products and familiarize the experimenter with the operation of recrystallization and vacuum filtration.

（3）To learn the identification of the purity of the solid product by the determination of the melting points or some specific chemical reaction.

【Principles】

Aspirin, also known as acetylsalicylic acid, is the most widely sold over-the-counter drug. It has the ability to reduce fever (an antipyretic), to reduce pain (an analgesic), and to reduce swelling, soreness and redness (an anti-inflammatory agent). Despite its side effects, aspirin remains the safest, cheapest and most effective non-prescription drug.

In this experiment, you will prepare aspirin by the reaction of salicylic acid with acetic anhydride, using concentrated sulfuric acid as a catalyst. The main side reaction is the dehydration reaction between the phenolic hydroxyl group and the carboxyl group. When the temperature is higher than 90℃, the intermolecular dehydration intensifies, resulting in water-insoluble polymers. The crude acetylsalicylic acid produced by the reaction contains unreacted salicylic acid and some by-products with high molecular weight. It can be purified by recrystallization with alcohol-water mixed solvent. Or dissolve acid in alkali solution first, remove insoluble matter by filtration and then acidify to make acetylsalicylic acid recrystallize and precipitate.

The main reaction:

$$\text{(salicylic acid: COOH, OH)} + (CH_3CO)_2O \xrightarrow{H_2SO_4} \text{(acetylsalicylic acid: COOH, OCOCH}_3\text{)} + CH_3COOH$$

Phenols form a colored complex with the ferric ion. The purple-red color indicates the presence of a phenol group. Note the aspirin no longer has the phenol group and thus it is one of the simplest and commonly used methods in laboratories to test the purity of acetylsalicylic acid by $FeCl_3$ solution. Measuring the melting point of it can also be used to test the purity.

【Apparatus and Materials】

Apparatus: Erlenmeyer flask, beaker, water-bath, measuring cylinder, test tube, magnetic stirring electric heating plate, melting point apparatus, vacuum filtration apparatus.

Materials: salicylic acid, acetic anhydride, concentrated sulfuric acid, concentrated hydrochloric acid, sodium bicarbonate, iron trichloride.

【Procedure】

1. Preparation

Prepare a water bath at 80-90℃ for use. Weigh 2.0 g of salicylic acid, 5 mL of acetic anhydride and place them in a dry 25 mL Erlenmeyer flask, and add 4 drops of concentrated sulfuric acid with constant swirling. Place the flask in the hot water bath for several minutes to dissolve solid material and wait with constant swirling for further 8-10 min to complete the reaction， and then pour the solution into a 100 mL beaker，add 30 mL ice-water to cool to induce crystallization，and collect the crystalline solid by vacuum filtration. Wash the solid twice with a small amount of distilled water（about 5 mL）.

2. Purification

Transfer your crude products to a 50 mL beaker and add 30 mL of saturated sodium bicarbonate solution. Swirl the solution to allow a complete neutralization between acids and sodium bicarbonate. After then， remove the insoluble substances by vacuum filtration. Pour all the filtrate to a clean beaker； add slowly concentrated hydrochloric acid into the solution with swirling the beaker until some crystalline solid comes out. Then place the beaker into an ice-water bath for complete crystallization. Collect the crystals by vacuum filtration. Dry the crystals under an infrared lamp. Weigh and calculate the yield.

3. Analysis of purity

（1）Label three test tubes： place a few crystals of salicylic acid into the first test tube， a few crystals of your product into the second test tube， and a small amount of commercial aspirin into the third test tube. Add 5 mL of distilled water to each test tube and swirl it to dissolve the crystals. Add 2-3 drops of 1% $FeCl_3$ to each test tube. Compare and record your observations.

（2）Place a few crystals of your product which is completely dried on a watch glass and which is ground into powder. Prepare 2 pieces of capillary melting tubes with your sample and determine the melting point by a melting point apparatus. The melting point of purified acetyl salicylic acid is 135-136℃.

3.10　二苄叉丙酮的制备

【实验目的】

（1）掌握微量有机合成实验方法。

（2）建立微量化学合成观念，培养绿色化学意识。

【实验原理】

绿色化学又称可持续化学，是指在化学和化学工程领域致力于从产品源头设计和生产中最大限度地减少有害物质的使用和产生。绿色化学聚焦于化学对环境的影响，包括防止污染和减少不可再生资源消耗的技术途径。世界上很多国家已把“绿色化学”作为新世纪化学发展的主要方向之一。

二苯乙烯基丙酮也称二苄叉丙酮，通常简写为 DBA，由于其结构的特殊性（见主反应式分子结构），它既能与亲核试剂发生反应，也能与亲电试剂发生反应，因此，在许多重要的有机合成反应中得到大量应用，如 Nazarov 环化、Michael 反应、Diels-Alder 反应等。经过近 80 多年的研究，人们已经合成出大量多取代二苯乙烯基丙酮类化合物。

本实验在碱性条件下，用过量的苯甲醛和丙酮交叉羟醛缩合生成目标物。

主反应式如下：

2PhCHO + (CH₃)₂C=O —NaOH, EtOH / 室温→ PhCH=CH–CO–CH=CHPh

【实验仪器与材料】

仪器：圆底烧瓶，烧杯，微型注射器，量筒，电磁搅拌加热板，减压过滤装置。

材料：无水乙醇，苯甲醛，丙酮，氢氧化钠，乙酸乙酯，pH 试纸。

【实验步骤】

用微型注射器（5 mL）在 25 mL 圆底烧瓶中先加入 5 mL 无水乙醇，再加入冰水浴后的 NaOH 溶液（NaOH，1 g；H_2O，7 mL），搅拌均匀，室温下用微型注射器（1 mL）加入苯甲醛（0.8 mL，8.2 mmol），磁力搅拌，然后向体系中缓慢滴加丙酮（0.3 mL，4 mmol），继续快速磁力搅拌，析出黄色固体，反应 5 min 后，将体系缓慢倒入 200 mL 冰水搅拌，析出大量固体，减压过滤，加 100 mL 水，水洗滤饼至中性。

把粗产物放入 5～10 mL 乙酸乙酯溶剂中，电加热板加热磁力搅拌至产物全部溶解，趁热减压热过滤，室温冰水冷却，析出黄色固体，再次减压过滤，产物干燥，称量，计算产率。

3.10 Preparation for Dibenzylideneacetone

【Objectives】

（1）To master the techniques of the micro organic synthesis.

（2）To establish the concept of micro chemical synthesis and increase the awareness of "green chemistry".

【Principles】

Green chemistry，also called sustainable chemistry，is an area of chemistry and chemical engineering focused on the designing of products and the processes that minimize the use and generation of hazardous substances. Green chemistry focuses on the environmental impact of chemistry，including technological approaches to preventing pollution and reducing consumption of nonrenewable resources. Many countries in the world have taken "green chemistry" as one of the main directions of chemical progress in the 21st century.

Because of its special structure（see the molecular structure in the main reaction formula），1, 5-diphenylpenta-1, 4-dien-3-one（DBA）can react with both nucleophilic reagents and electrophilic reagents. Therefore，it has been widely used in many important organic synthesis reactions，such as Nazarov cyclization reaction，Michael reaction and Diels-Alder reaction. After nearly 80 years of research，many polysubstituted stilbenyl acetone compounds have been synthesized.

In this experiment，we will learn an efficient protocol for crossed aldol condensation of ketones with acetone using excess quantities of benzaldehyde under alkalic condition.

The main reaction：

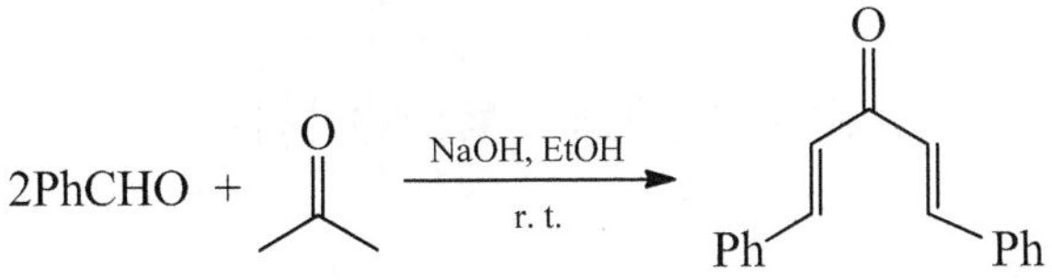

2PhCHO + O (acetone) $\xrightarrow[\text{r. t.}]{\text{NaOH, EtOH}}$ Ph, O, Ph (dibenzylideneacetone)

【Apparatus and Materials】

Apparatus：round-bottom flask，beaker，micro-syringe，measuring cylinder，magnetic stirring electric heating plate，vacuum filtration apparatus.

Materials：anhydrous ethanol，benzaldehyde，acetone，sodium hydroxide，ethyl acetate，pH test paper.

【Procedure】

Add 5 mL of anhydrous ethanol into a 25 mL round-bottom flask with a micro-syringe（5 mL），and then the NaOH solution（NaOH，1 g；H_2O，7 mL）after ice-water bath is added. Stir to make them mixed evenly. Under room temperature，benzaldehyde（0.8 mL，8.2 mmol）is added into the system with a micro-syringe（1 mL），and stirred by magnetic force. Then acetone（0.3 mL，4 mmol）is slowly dripped into the system. Continue rapid magnetic stirring，and precipitate yellow solids. After 5 minutes，the system is slowly poured into 200 mL ice water to stir，precipitate a large number of solids，the product is collected by vacuum filtration，add 100 mL water to wash until the washings are neutral to litmus.

The crude product is dried under vacuum and then recrystallize from EtOAc（5-10 mL）. Dry，weigh and calculate the yield.

3.11　纳迪克酸酐的制备

【实验目的】

（1）学习 Diels-Alder 反应原理。

（2）掌握 Diels-Alder 反应合成实验基本操作。

【实验原理】

Diels-Alder（第尔斯-阿尔德）反应又名双烯加成，由共轭双烯与烯烃或炔烃反应生成六元环状化合物。Diels-Alder 反应具有丰富的立体选择性、立体专一性和区域选择性等特点。

在合成有机化学中，由非环前体生成碳环的反应具有重要意义，由于 Diels-Alder 反应中两个新的碳碳键和一个新六元环是一步形成，因此它已成为制备复杂分子的关键步骤，可大大减少反应步骤，提高合成效率，是有机化学合成反应中非常重要的碳碳键形成手段之一，也是现代有机合成中常用反应之一。

1928 年，德国化学家第尔斯和他的学生阿尔德首次发现并报道了这种新型反应，他们也因此获得 1950 年诺贝尔化学奖。

本实验将会重现 1928 年 Diels-Alder 反应原始文文献中马来酸酐（一个经典的 Diels-Alder 二烯体）与二烯（三个共轭单萜之一）反应。

主反应式如下：

O O O O O O

此加成产物还具有双键，能使高锰酸钾溶液或溴的四氯化碳溶液褪色。

【实验仪器与材料】

仪器：锥形瓶，水浴锅，烧杯，量筒，电磁搅拌加热板，减压过滤装置。

材料：顺丁烯二酸酐（马来酸酐），环戊二烯，石油醚（60～90℃），乙酸乙酯。

【实验步骤】

在 50 mL 锥形瓶中加入 3 g 马来酸酐和 10 mL 乙酸乙酯，用热水浴加热使固体物全部溶解，然后加入 10 mL 石油醚（60～90℃）。用冰水浴冷却（这时可能会有少量沉淀析出，但不会影响反应），再加入 2.4 g（3 mL）新蒸馏的环戊二烯。将盛有反应液的锥形瓶置于冰水浴中并不断摇动，直到白色固体析出，放热停止。用水浴加热使析出的固体全部溶解，然后静置使其缓缓地冷却，得到白色针状结晶，减压过滤，用 5 mL 乙酸乙酯和石油醚混合液（体积比为 1∶1）淋洗，得产物纳迪克酸酐（顺式-3，6-内亚甲基-1，2，3，6-四氢邻苯二甲酸酐），干燥，称量，计算产率。

3.11 Preparation for Nadic Anhydride

【Objectives】

（1）To learn the principle of the Diels-Alder reaction.

（2）To master the organic synthetic techniques of the Diels-Alder reaction.

【Principles】

Diels-Alder reaction，also known as diene addition，is that alkene and alkyne undergo a unique cycloaddition reaction with conjugated diene to form the hexahydroxy compound. Diels-Alder reactions are characterized by abundant stereoselectivity，stereospecificity and regioselectivity.

In synthetic organic chemistry，the reaction of carbon ring formation from non-cyclic precursors is of great significance. Because the two new carbon-carbon bonds and a new six-membered ring are formed in one step in Diels-Alder reaction，this has become a key step in the preparation of complex molecules，which can greatly reduce the reaction steps and improve the synthesis efficiency. Diels-Alder reaction is a very important means of carbon-carbon bond formation in organic chemical synthesis reaction and also one of the common reactions in modern organic synthesis.

Otto Paul Hermann Diels and his student Kurt Alder，two German chemists，published their first paper on this reaction in 1928. Twenty-two years later，the reaction was so important in organic synthesis that these two men shared the Nobel Prize in chemistry.

In this experiment，you will examine the reaction of maleic anhydride，a classic Diels-Alder dienophile，with the diene，one of the three conjugated monoterpenes，was reported in the original paper by Diels-Alder in 1928.

The main reaction：

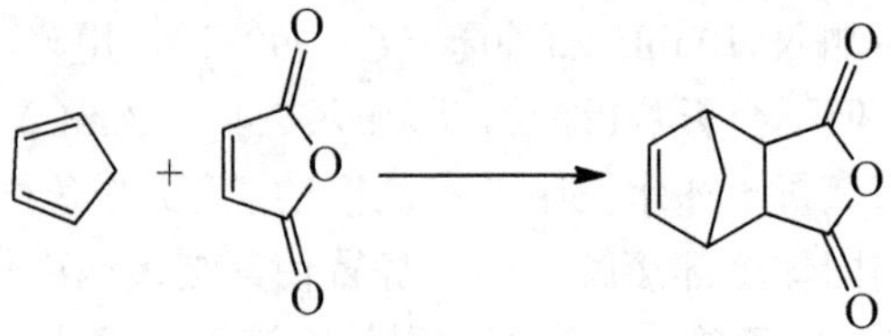

This adduct has retained a carbon-carbon double bond，so it can make the solution of potassium permanganate or the solution of bromine in carbon tetrachloride depigmentize.

【Apparatus and Materials】

Apparatus：Erlenmeyer flask，water-bath，beaker，measuring cylinder，magnetic stirring electric heating plate， vacuum filtration apparatus.

Materials：maleic anhydride，cyclopentadiene，petroleum ether（60-90℃），ethyl acetate.

【Procedure】

Place 3 g of maleic anhydride in a 50 mL Erlenmeyer flask and dissolve it in 10 mL of ethyl acetate by warming in a hot water bath. Add 10 mL of petroleum ether（60-90℃）and then cool the solution in an ice bath. Add 2.4 g（3 mL）of freshly distilled cyclopentadiene to the maleic anhydride solution and swirl the solution to mix in the ice-water bath until the white product crystallizes from solution and the initial exothermic reaction is over. Heating the flask in a water bath until the product has dissolved，and then allowing the solution to cool slowly and stand undisturbed. The white needle-shape crystals are obtained. Collect the product Nadic anhydride by vacuum filtration and wash it with 5 mL the mixed solvent of ethyl acetate and petroleum ether（$V:V=1:1$）. Afterward dry，weigh and calculate the yield.

3.12　无水乙醇的制备

【实验目的】

（1）学习制备无水乙醇的原理。
（2）掌握回流、蒸馏基本操作。

【实验原理】

有机合成化学实验中经常需要高纯度的乙醇，即“绝对乙醇”。市售乙醇纯度多为 95.5%，常含有 4.5%水，由于乙醇与水易形成恒沸点为 78.5℃的混合物，所以不能通过蒸馏直接提纯。在实验室中要得到纯度较高的乙醇，常用加入氧化钙加热回流，使乙醇中的水与氧化钙作用，生成不挥发性氢氧化钙来除去水分。这样得到的无水乙醇纯度较高，可达 99.5%。要制备纯度更高的绝对乙醇，可用金属镁进行处理。

【实验仪器与材料】

仪器：圆底烧瓶，锥形瓶，水浴锅，烧杯，量筒，干燥管，温度计，接液管，电磁搅拌加热套，回流装置，蒸馏装置。

材料：95%工业乙醇，氧化钙，氯化钙。

【实验步骤】

在 200 mL 圆底烧瓶中，加入 100 mL 95%工业乙醇和 25 g 氧化钙，再加入磁力搅拌子，装上回流冷凝管，顶端放置内有氯化钙的干燥管，在水浴上回流加热 60 min。

冷却后取下冷凝管，换成蒸馏装置，电热套加热蒸馏，调整加热功率，保持蒸馏速度为每秒一到两滴，并记录第一滴蒸馏液的温度。当温度升至 77℃并保持稳定时，用干燥的量筒替换接收器，收集 77～79℃范围内的蒸馏液，至无液滴流出时停止蒸馏，需特别注意烧瓶内不要蒸馏至干，烧瓶中应留少量液体（通常为 0.5～1mL），防治局部过热爆炸。量取体积，计算回收率。

3.12　Preparation for Absolute Ethanol

【Objectives】

（1）To learn the principles of preparing absolute ethanol.

（2）To master the procedures of reflux and simple distillation.

【Principles】

Ethanol with a high degree of purity，which is called "absolute ethanol，" is frequently required in organic synthetic chemistry. But it cannot be prepared by distillation，owing to the formation of an azeotrope which the azeotropic boiling point with 95.5 percent of ethanol and 4.5 percent of water is 78.5℃. For some purposes absolute ethanol of about 99.5% purity is satisfactory，which is commercially available or conveniently prepared by the dehydration of rectified spirit with calcium oxide in the lab. Calcium oxide is added into the rectified spirit，and the mixture is heated to reflux，which results in the reaction of calcium oxide with water. In this process，the involatile calcium hydroxide is produced. For obtaining the higher-purity absolute ethanol，magnesium metal can be used.

【Apparatus and Materials】

Apparatus：round bottom flask，Erlenmeyer flask，water bath，beaker，measuring cylinder，drying tube，thermometer，distillation adapter，magnetic stirring electric heating mantle，reflux apparatus，simple distillation apparatus.

Materials：95% industrial ethanol，calcium oxide，calcium chloride.

【Procedure】

Pour 100 mL of 95% industrial alcohol and calcium oxide（25 g）into a 200 mL round bottom flask. Fit the flask with a ball condenser and place a stir bar in the flask，and then a calcium chloride drying tube. Reflux the mixture gently for 60 min in water bath.

Stop heating，change to the distillation apparatus after cooling down. Assemble a simple distillation apparatus. Adjust the heating power，maintain the distillation at a rate of one or two drops per second，and record the temperature of the first drop of the distillate. When the temperature rises to 77℃ and remains stable，replace the receiver with a dry measuring cylinder to collect the distillate in the range of 77-79℃. If no more distillates to be collected，cease the distillation. Great care should be taken not to distill the solution to dryness because，in some cases，high-boiling explosive peroxides can become concentrated. Small volume of the liquid（usually 0.5-1 mL）should be remained in the flask. Measure the collected absolute ethanol and calculate the yield.

第三部分　有机合成新方法
Part 3　New Methods of Organic Synthesis

3.13　微波辅助合成反应——二苯乙炔的合成

【实验目的】

（1）学习微波辅助合成基本原理。

（2）掌握微波辅助合成实验基本操作技术。

【实验原理】

在合成化学中使用微波加热是一种较新方法。在 1986 年，首次使用微波加热有机合成反应就已被报道，但直到几年后，这种方法才被合成化学家广泛使用。利用微波技术作为热源不仅大大减少了有机反应的加热时间，而且与采用传导加热方法相比，反应效率更高、选择性更好。在合成反应中使用微波加热的另一个优点是与传统加热方法相比，所需溶剂更少。此外，在微波装置中加热反应不需要特殊容器。

在本实验中，1,2-二溴-1,2-二苯基乙烷在微波辐射条件下脱溴化氢生成二苯乙炔。通常在没有采用特殊装置的情况下，实现该反应所需的温度接近 200℃，所以必须使用高沸点溶剂，如三乙二醇(TEG)。然而，如果利用微波合成仪进行该反应，反应温度可以下降到 150℃，甲醇为溶剂，同时缩短反应时间。

主反应式如下：

H　Br

H　Br

+ 2KOH $\xrightarrow[\text{Microwave Radiation}]{CH_3OH,\ 150℃}$

【实验仪器与材料】

仪器：烧杯，量筒，滤纸，带盖耐压管，磁力搅拌子，微波合成仪，磁力搅拌装置，减压过滤装置。

材料：1,2-二溴-1,2-二苯基乙烷，氢氧化钾，甲醇，乙醇。

【实验步骤】

在 10 mL 耐压管中加入磁力搅拌子、 0.40 g 1,2-二溴-1,2-二苯基乙烷、 0.2 g

固体氢氧化钾和 2 mL 甲醇。 盖上耐压管帽，轻轻摇动或放置在磁力搅拌器上使反应体系充分混匀。将耐压管放置在微波合成仪中。

将微波合成仪反应温度设定为 150℃，功率为 25 W，辐射时间为 5 min，保温时间为 5 min。

从微波合成仪中取出耐压管冷却至室温，加入 5 mL 水，放入冰水浴中 5 min。减压过滤收集沉淀粗产物，用约 2 mL 冷水洗涤固体。

用少量 95%乙醇或乙醇-水混合物重结晶产物。如果重结晶溶液被缓慢冷却，静置，则会得到粒大、无色透明晶体。

3.13 Microwave-Assisted Synthesis Reaction—Synthesis of Diphenylacetylene

【Objectives】

（1）To learn the principles of the microwave-assisted synthesis reaction.

（2）To master the technique of the microwave-assisted synthesis reaction.

【Principles】

Using microwave heating in synthetic chemistry is a relatively new methodology. Although the first reports of the heating organic reactions using microwave appeared in 1986，it was not until several years later that this method became widely used by synthetic chemists.The use of microwave technology as heat sources drastically reduces the heating time of organic reactions，and the reactions often proceed more efficiently and selectively than when conduction heating methods are used. Another benefit of using microwave heating in synthesis is that less solvent is required than when using traditional heating methods. Furthermore，reactions heated in a microwave instrument do not necessarily require special apparatus.

In this experiment，you will perform the dehydrobromination of *meso*-diphenylethane dibromide to give diphenylacetylene according to equation. Because the temperature required to effect elimination of hydrogen bromide from the intermediate vinyl bromide is nearly 200℃，it is necessary to use a high-boiling solvent such as triethylene glycol（TEG）so the reaction may be carried out without special apparatus. However，if a microwave apparatus is available for heating，the reaction can be effected in a pressure-rated tube at 150℃ using methanol as solvent and by shortening the reaction time.

Main reaction:

【 Apparatus and Materials 】

Apparatus: beaker, measuring cylinder, filter paper, pressure-rated tube with cap, stir bar, microwave synthesis apparatus, magnetic stirring apparatus, vacuum filtration apparatus.

Materials: *meso*-diphenylethane dibromide, potassium hydroxide, methanol, ethanol.

【 Procedure 】

Prepare a 10 mL pressure-rated tube with the stir bar and add 0.40 g of *meso*-diphenylethane dibromide, 0.2 g of solid potassium hydroxide, and 2 mL of methanol. Cap the pressure-rated tube and gently shake it or place it on a magnetic stirrer to facilitate the initial mixing of its contents. Place the tube in the cavity of the microwave apparatus.

The reaction temperature should be set at 150℃ and the power set at a maximum of 25 W with 5 min ramp time and 5 min hold time.

Allow the mixture to cool to room temperature and remove the tube from the microwave apparatus. Add 5 mL of water and place the tube in an ice-water bath for 5 min. Collect the diphenylacetylene that precipitates by vacuum filtration. Wash the solid with about 2 mL of cold water.

Recrystallize the product from a small quantity of 95% ethanol or an ethanol-water mixture. If the solution is allowed to cool slowly undisturbed, you will obtain large, colorless crystals.

3.14　超声波辐射反应——苯亚甲基苯乙酮的合成

【实验目的】

（1）学习超声波辐射反应基本原理。

（2）掌握超声波辐射反应实验基本操作技术。

【实验原理】

超声波作为活化和促进化学反应的高新技术在 20 世纪 80 年代中期以后才发展起来，这一技术在化学化工中的应用研究形成了一门新兴的交叉学科——超声化学，超声化学的研究和发展已开始使超声波技术从实验室走向工业化。通常把频率范围为 20～1000 kHz 的声波称为超声波。有机化学合成所用的超声波频率一般为 20～80 kHz，合成化学主要利用超声波的声空化效应，在介质的微区和极短的时间内产生高温高压的高能环境，并伴有强大的冲击波和微射流，以及放电、发光等，这就为促进和启动化学反应创造了一个极端的物理环境。超声波不仅能增加反应速率，易于引发反应，降低苛刻的反应条件，而且可以改变反应的途径和选择性。因此，这一技术一经用于具有重要经济价值的反应即显示出巨大的应用前景。

主反应式如下：

$$\text{PhCOCH}_3 + \text{PhCHO} \xrightarrow[25\sim30^\circ\text{C}]{\text{NaOH, EtOH}} \text{PhCOCH=CHPh} + H_2O$$

【实验仪器与材料】

仪器：超声波清洗器，锥形瓶，量筒，磁力搅拌装置，减压过滤装置。

材料：苯乙酮，苯甲醛，氢氧化钠，95%乙醇。

【实验步骤】

在 50 mL 锥形瓶中，依次加入 2.1 mL 10% 氢氧化钠水溶液，2.5 mL 95%乙醇，1 mL 苯乙酮，摇匀，冷却至室温，再加入 0.8 mL 新蒸馏过的苯甲醛。将反应瓶置于超声波清洗槽中，并使清洗槽中水面略高于反应瓶中的液面，控制清洗槽中水温在 25～30℃，启动超声波清洗器，超声波辐射 30～35 min，停止反应。然后将反应瓶置于冰水浴中冷却，使其结晶完全。减压过滤，用少量冰水洗涤产品至滤液呈中性。粗产品用 95%乙醇重结晶。干燥，称量，计算产率。

3.14　Ultrasonic Radiation Reaction—Synthesis of Benzyl Acetophenone

【Objectives】

（1）To learn the principles of the ultrasonic radiation reaction.

（2）To master the techniques of the ultrasonic radiation reaction.

【Principles】

Ultrasonic wave，a new technology to activate and promote chemical reactions，has been developed after the middle period of the 1980s. This technology，which is applied and researched in the field of chemistry and chemical engineering，has formed an emerging cross-discipline—ultrasonic chemistry. Because of the research and development of the ultrasonic chemistry，this technology has begun to move from the laboratory to the industrialization. The sound wave whose frequency is in the range of 20-1000 kHz is called the ultrasonic wave. The used ultrasonic wave frequency in organic synthesis is usually at 20-80 kHz. The synthetic chemistry mainly uses the cavitation effect of the ultrasonic wave. A high energy environment of high temperature and high pressure is produced in the micro-area of medium and the extremely short time，accompanied by the powerful shock wave and the micro-jet flow，as well as the discharge and radiation，etc. It creates an extreme physical environment for promoting and opening the pass of chemical reactions. Ultrasonic wave can not only increase the reaction rate，initiate the reaction easily，reduce the harsh reaction conditions，but also can change the reaction pathway and selectivity. Once this technology is applied to the reaction with important economic value，it will produce huge application prospect.

The main reaction：

$$C_6H_5COCH_3 + C_6H_5CHO \xrightarrow[25\sim30℃]{NaOH,\ EtOH} C_6H_5COCH{=}CHC_6H_5 + H_2O$$

【Apparatus and Materials】

Apparatus：ultrasonic cleaner，Erlenmeyer flask，measuring cylinder，magnetic stirring apparatus，vacuum filtration apparatus.

Materials：acetophenone，benzaldehyde，sodium hydroxide，95% ethanol.

【Procedure】

In a 50 mL Erlenmeyer flask place 2.1 mL of 10% sodium hydroxide solution, 2.5 mL of 95% ethanol, 1 mL of acetophenone. Cool the mixture to room temperature, shake well, and add 0.8 mL of freshly redistilled benzaldehyde. Fix the flask in the ultrasonic wave cleaning instrument whose trough has been added some water in advance. The water surface in the cleaning trough should be slightly higher than the liquid level in the flask. Turn on the ultrasonic wave cleaning instrument, and radiate the mixture in the flask by ultrasonic wave for about 30-35 min. Control the temperature of the water in the trough at 25-30℃ during the radiation. Then chill the reaction mixture in an ice-water bath for thirty minutes or longer to crystallize completely. Collect the product with vacuum filtration and wash it thoroughly with ice water, until the washings are neutral to pH indicator paper. Recrystallize the crude product from 95% ethanol. After drying, weigh and calculate the yield.

3.15　有机光化学反应——苯频哪醇的合成

【实验目的】

（1）学习有机光化学反应基本原理。

（2）掌握有机光化学反应实验基本操作技术。

【实验原理】

由光激发分子导致的化学反应称为光化学反应。通常紫外光和可见光能引起光化学反应，其波长在 200～700 nm 范围内。能发生光化学反应的物质一般具有不饱和键，如烯烃、醛、酮等。

二苯甲酮的光化学还原是研究最为详细的光化学反应。将二苯甲酮溶于氢给予体的溶剂（如异丙醇）中，在紫外光照射下，生成二聚体苯频哪醇。实验证明，二苯甲酮的光化学还原是二苯甲酮的 $n \to \pi^*$ 三线态（T_1）的反应。二苯甲酮的异丙醇溶液用 300～350 nm 的紫外光照射时，异丙醇不吸收光能，只有二苯甲酮的羰基接受光能后，外层的非键电子发生 $n \to \pi^*$ 跃迁，经单线态（S_1）、系间窜跃成三线态，由于三线态有较长的半衰期和较多的能量（314～334.7 $kJ \cdot mol^{-1}$），可以从异丙醇的 C2 上夺取氢，使 C2 上的 C—H 键均裂，各自形成自由基，再经自由基的转移、偶合形成苯频哪醇。

主反应式如下：

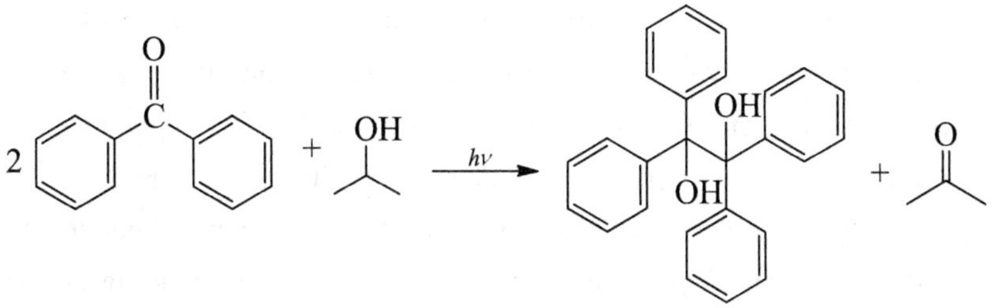

【实验仪器与材料】

仪器：烧杯，量筒，试管，熔点仪，磁力搅拌装置，减压过滤装置。

材料：二苯甲酮，异丙醇，冰醋酸。

【实验步骤】

在一支 10 mL 试管中加入 1 g 二苯甲酮和 5 mL 异丙醇，在温水浴中加热，使二苯甲酮溶解。向试管中加入 2 滴冰醋酸，充分振摇后再补加异丙醇至试管口，以使反应尽量在无空气条件下进行。用塞子将试管塞住，置试管于烧杯中，并放在光照良好的窗台上，光照一周，试管内有大量无色晶体析出。经减压过滤、干燥后即得苯频哪醇粗品。粗产物可用冰醋酸作溶剂进行重结晶。干燥，称量，计算产率，测熔点，验证纯度。

3.15 Organic Photochemical Reaction—Synthesis of Benzopinacol

【Objectives】

（1）To learn the principles of the organic photochemical reaction.

（2）To master the techniques of the organic photochemical reaction.

【Principles】

The chemical reaction，which can occur due to the molecules excited by light，is called photochemical reaction. Generally，ultraviolet light and visible light can cause the photochemical reaction，and the wavelength of light is in the range of 200-700 nm. The substances which contain unsaturated bonds，such as alkenes，aldehydes，ketones，etc.，can generally carry on the photochemical reactions.

The photoreduction of benzophenone is one of the most thoroughly studied photochemical reactions. If benzophenone is dissolved in a "hydrogen-donor" solvent，such as propanol，and exposed to ultraviolet light，an insoluble dimeric product，benzopinacol，will form. Experiments show that the photoreduction of benzophenone is a reaction of the $n \rightarrow \pi^*$ triplet state（T_1）of benzophenone. When the 2-propanol solution of benzophenone is exposed to ultraviolet light in the range of 300-350 nm，only the π electron of the carbonyl π bond in benzophenone can absorb the photon of energy，but 2-propanol can not absorb it. A non-bonding electron from the highest-energy occupied orbital in the carbonyl bond happens the $n \rightarrow \pi^*$ transition to the triplet state via the singlet state（S_1）and intersystem crossing. Because the triplet state has a longer half life period and considerable energy（314-334.7 $kJ \cdot mol^{-1}$），it can abstract a hydrogen atom on the sec-carbon from 2-propanol. Then the C—H bond on the sec-carbon happens homolytic cleavage. The diphenylhydroxymethyl radicals and isopropyl radicals are formed. Two of diphenylhydroxymethyl radicals，once formed，may couple to form benzopinacol.

The main reaction：

$$2\,PhC(=O)Ph + (CH_3)_2CHOH \xrightarrow{h\nu} Ph_2C(OH)C(OH)Ph_2 + CH_3COCH_3$$

【Apparatus and Materials】

Apparatus: beaker, measuring cylinder, test tube, melting point apparatus, magnetic stirring apparatus, vacuum filtration apparatus.

Materials: benzophenone, isopropanol, glacial acetic acid.

【Procedure】

Add 1 g of benzophenone and 5 mL of isopropanol to a 10 mL test tube, dissolve the solid in a warm water bath. Add two drops of glacial acetic acid to the tube, shake thoroughly, and then add sufficient isopropanol to fill the tube in order to exclude the air in the tube. Stopper the test tube tightly with a rubber stopper. Place it in a beaker on a window sill where they will receive direct sunlight for a week. A large amount of colorless crystals will be separated from the solution. Collect the colorless crystals with vacuum filtration and the crude product from glacial acetic acid. After drying the product, weigh, and calculate the yield, and determine its melting point.

3.16　有机电化学反应——碘仿的合成

【实验目的】

（1）学习有机电化学反应基本原理。

（2）掌握有机电化学反应实验基本操作技术。

【实验原理】

有机电化学合成是利用电解反应来合成有机化合物。有机电化学合成是绿色化学中有机合成洁净技术的重要组成部分，电化学合成法与常规合成法相比有以下特点：①可自动控制；②反应条件温和，能量效率高；③环境相容性高；④性价比高。所需设备简单，操作费用低。设计合理的电解池结构，利用先进电极材料，可达到零排放的要求。目前，电化学方法在有机合成上的应用已引起人们的广泛关注，日益被化学、化工界所重视。

碘仿又称黄碘，为亮黄色晶体，在医药和生物化学中作防腐剂和消毒剂。

【实验仪器与材料】

仪器：有机玻璃板，$1^{\#}$电池，烧杯，量筒，熔点仪，磁力搅拌装置，减压过滤装置。

材料：碘化钾，丙酮，乙醇。

【实验步骤】

用 50 mL 的小烧杯作电解槽，两支 $1^{\#}$电池碳棒作电极，把它们垂直固定在硬纸板或有机玻璃板上。烧杯中加 40 mL 水、2.2 g 碘化钾和 0.5 mL 丙酮。将烧杯放置在电磁搅拌器上，开动电磁搅拌器使药品溶解，接通电源（6 V 直流电），在室温下电解。随着反应的进行，有黄色沉淀生成。反应 60 min 左右转换电极，以加快反应速率。反应 150 min 后，切断电源，停止反应，再继续搅拌 1～2 min，然后减压过滤，滤饼用少量水洗涤两次，空气中自然干燥后即得粗产品。粗产品用乙醇重结晶后得到纯品，产品经晾干后，称量，计算产率，测熔点，验证纯度。

3.16　Organic Electrochemical Reaction—Synthesis of Iodoform

【Objectives】

（1）To learn the principles of the organic electrochemical reaction.

（2）To master the techniques of the organic electrochemical reaction.

【Principles】

Electroorganic synthesis is a way to synthesize organic compounds by electrolytic reaction. It is an important component of clean technology on organic synthesis in green chemistry. Compared with conventional methods，it has such characters：①it can autocontrol；②the reaction conditions are mild and the energy efficiency is high; ③the environmental compatibility is high; ④it is very economical. The required equipment is uncomplex and the cost of equipment is lower. The electrolytic cell structure of rational design and the use of advanced electrode materials can achieve the requirements of zero emission. At present，the application of electrochemical method in organic synthesis has attracted wide attention. The electrochemistry method has been recognized increasingly in the field of chemistry and chemical industry.

Iodoform，also called yellow iodine，is a bright yellow crystal. It can be used as preservatives and antiseptics.

【Apparatus and Materials】

Apparatus：perspex sheet，$1^{\#}$ battery，beaker，measuring cylinder，melting point apparatus，magnetic stirring apparatus，vacuum filtration apparatus.

Materials：potassium iodide，acetone，ethanol.

【Procedure】

Make use of a 50 mL beaker as electrolyzer and two carbon sticks of $1^{\#}$ battery as electrodes. The electrodes are fixed vertically on a hard cardboard or a plexiglass plate. Add 40 mL of water，2.2 g of potassium iodide and 0.5 mL of acetone to the beaker. Stir the mixture until the solution is clear. Switch on the power source of electrodes（connect 6 V direct current）and electrolyze at room temperature. A great amount of yellow precipitation is produced during the reaction. Change the electrodes to stop the reaction after about 60 min to accelerate the reaction. Switch off the power of electrodes to stop the reaction after 150 min，and continue stirring for 1-2 min again. Collect the yellow precipitate of iodoform with suction filtration and wash it with a little cold water twice，dry it in the air. Recrystallize the crude product from ethanol. After drying，weigh and calculate the yield，determine its melting point.

第四部分 有机化合物物理常数的测定及官能团鉴定
Part 4 Identification of Organic Compounds by Determination of Physical Constants and Characteristic Reactions

3.17 固体有机物熔点测定

【实验目的】

（1）学习熔点测定的原理和意义。

（2）掌握微量熔点仪的操作。

【实验原理】

熔点是指该物质的液相和固相达到平衡时的温度。因此，通常报告化合物的熔点范围，较低的温度是液体的第一个微滴出现的温度，较高的温度是固体完全熔化的温度。

纯净的固体化合物一般都有固定熔点，固-液两相之间的变化非常敏锐，从初熔到全熔的温度范围（称为熔距或熔程）一般不超过0.5～1℃（除液晶外）。当其中混有杂质时，熔点就有显著的变化，将会使其熔点下降且熔程变长。因此，准确测定晶体样品的熔点，将测得的数据与文献记载的标准数据相比较，如果相符，则说明样品是纯净的，如果低于文献值，则说明样品不纯净。

有许多不同类型的仪器可用于测定有机物的熔点。两种主要测定熔点方法是提勒（Thiele）管法和显微熔点仪法。

1. Thiele 管法测定熔点

样品在一个特殊装置中缓慢加热，该装置是b形管，称为Thiele管，配有温度计和加热液体作为加热槽。常用的热浴液体油是液状石蜡油（b.p. 220℃）或硅油（b.p. 250℃）。

在测定时注意两个温度：第一个是第一滴液体在晶体中形成的点，第二个是晶体整个变成透明液体的点。记录并给出这一熔化范围。手册中的一些熔点数据是实际实验室数据的平均值。

先把样品装入熔点管中。熔点管是一端开口、另一端封口的合适孔径的毛细管。将干燥的固体试样在表面皿上堆成小堆，将熔点管的开口端插入试样中，装入少量待测试样。然后把熔点管竖立起来，开口端朝上，将一较大口径的玻璃管（长70～80 cm）垂直于桌面放置，使熔点管在玻璃管内自由落下，重复几次，使样品装入熔点管底，样品高度为 2～3 mm 即可。为使测定结果准确，样品要研得极细，填充要均匀紧密。

在图 3.3 中所示在 Thiele 管侧臂弯曲位置加热。载热体被加热后在管内对流循环，使温度变化比较均匀。在测定已知熔点的样品时，可先以较快速度加热，在距离熔点 10～15℃时，应控制加热强度使温度计指数以每分钟 1～2℃的速度上升，直到熔化，测出熔程。加热速度过快会使被测物熔程较大而得不到准确的结果。在测定未知熔点样品时，应先粗测熔点范围，再按前述方法精测。测定时，应观察和记录样品开始塌落仅有液相产生时（初熔）和固体完全消失时（全熔）的温度读数，所得数据即为该物质的熔程。还要观察和记录在加热过程中是否有萎缩、变色、发泡、升华及炭化等现象，以供分析参考。熔点测定至少要有两次重复数据，否则需再次测定。每次测定要用新毛细管重新装入新样品测定，不可将前面测定后留下的凝固物再次测量使用。

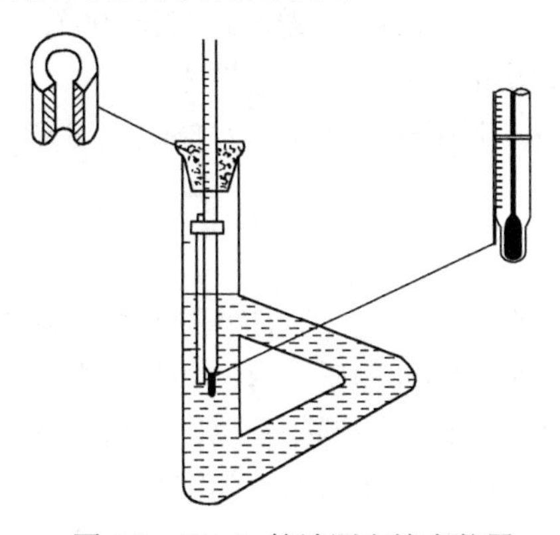

图 3.3　Thiele 管法测定熔点装置

2. 显微熔点仪测定熔点

显微熔点仪型号较多，其外形如图 3.4 所示，因有半自动和全自动的不同而操作有所差异，但共同特点是使用样品量少（2～3 颗小结晶），可通过显微镜直接观察晶体在加热过程中变化的全过程，如结晶的失水、多晶的变化及分解，能测量样品的熔点范围由 20℃至 300℃。在使用这种仪器前必须仔细阅读仪器使用指南，严格按操作规程进行。

3. 温度计校正

测定熔点时，温度计上显示的熔点与真实熔点之间常有一定的偏差，为了测得精确的熔点，通常需要先对温度计进行校正。校正包括：在温度计的量程范围内，测量一系列标准物质的熔点，以测定值与标准值之间的差值来消除用此温度计实际测量时的偏差。

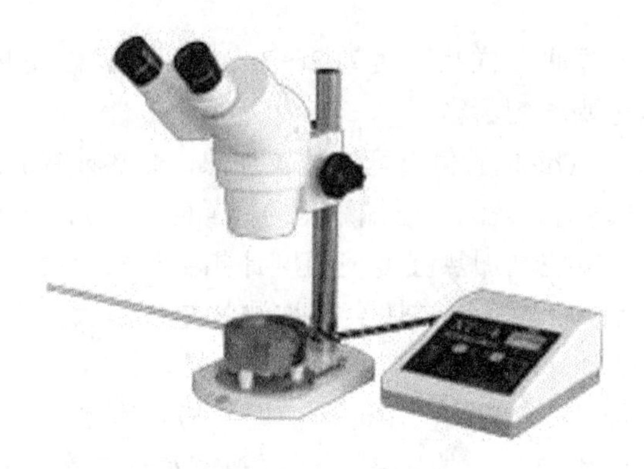

图 3.4　显微熔点仪

【实验仪器与材料】

仪器：提勒管（b 形管），毛细管，温度计（≥300℃），带缺口的橡皮塞，长玻璃管（70～80cm），表面皿，橡皮筋，熔点仪。

材料：液状石蜡，萘，苯甲酸，尿素，水杨酸，对苯二酚。

【实验步骤】

从提供的已知熔点化合物中，选一到两种化合物，用提勒管装置测定熔点。粗测 1 次，精测 2 次，记录实测的物质熔程，并与标准值比照，对比分析结果。

从已测过熔点的化合物中任选一种化合物，分别掺杂 5%和 10%另一种物质。混合均匀，测定样品熔点，验证杂质对熔点的影响。

精确测定由指导教师提供的未知样品的熔点。若实验条件允许，用提勒管装置和显微熔点仪分别测定相同样品的熔点，对比结果。

3.17　Determining the Melting Point of Organic Solids

【Objectives】

(1) To learn the identification of organic solids and their purity by determining the melting points.

(2) To master the operations of the microscopic melting point meter.

【Principles】

The melting point is the temperature at which solid and liquid exist together in equilibrium. Accordingly，a melting-point range of a compound is typically reported with the lower temperature at which the first tiny drop of liquid appears and the higher temperature at which the solid has completely melted.

The melting point is actually determined as a melting range. The melting points of pure substances occur over a very narrow range. Generally，the temperature range from primary melting to full melting（melting distance or melting range）does not exceed 0.5-1℃（except liquid crystal）. If the amount of impurity is increased，the vapor pressure of the liquid is lowered even further and therefore the melting point is lowered more. Hence，the criteria for purity of a solid are the narrowness of the melting point range and the correspondence to the value found in the literature. The broadening of the melting range that results from introducing an impurity into a pure compound may be used for identifying if a substance is pure or not.

Many different types of instruments are available to determine melting points. The two principal types of the melting point apparatus are the Thiele tube and microscopic melting point meter.

1. Measurement of melting point by Thiele tube

The sample is heated slowly in a special apparatus，which is a “b” -shaped tube called Thiele tube，equipped with a thermometer and heating liquid as a heating bath. The commonly used heating bath liquid oil is liquid paraffin wax oil（b.p. 220℃）or silicone oil（b.p. 250℃）.

In determining，two temperatures are noted. The first is the point at which the first drop of liquid forms among the crystals，and the second is the point at which the whole mass of crystals turns to a clear liquid. The melting point is then recorded giving this range of melting. Note that some melting point data in handbooks are the average of the actual lab figures.

The first step in determining a melting point is transferring the sample into a melting point capillary tube. Such tubes have one sealed end. The proper method for loading the sample into the capillary tube is as follows. Place a small amount of the drying solid on a clean watch glass and press the open end of the tube into the solid to force a small amount of solid into the tube. Then take a piece of glass tubing about 70-80 cm in length，hold it vertically on a hard surface such as the bench top，and drop the capillary tube down the larger tubing several times with the sealed end down. This packs the solid sample (about 2-3 mm in height) at the closed end of the capillary tube. In order to make the results accurate，the samples should be ground very finely and filled uniformly and tightly.

A simple type of the melting-point apparatus is the Thiele tube（see Figure 3.3）. On heating the bent side-arm，the heated liquid circulates and raises the temperature of the sample in such a way that no stirring of the bath liquid is required. The melting point apparatus is heated comparatively rapidly with a small flame until the temperature of the bath is within 10-15℃ of the melting point of the substance，and then slowly and regularly at the rate of about 1-2℃ per minute until the compound melts completely. The temperature at which the substance commences to liquefy and the temperature at which the solid has disappeared are observed，i.e. the melting point range. The determination of melting points should be repeated at least twice, otherwise it should be determined again. New capillaries should be used to reload the new samples for each determination，and the coagulants left behind after the previous determination should not be re-measured.

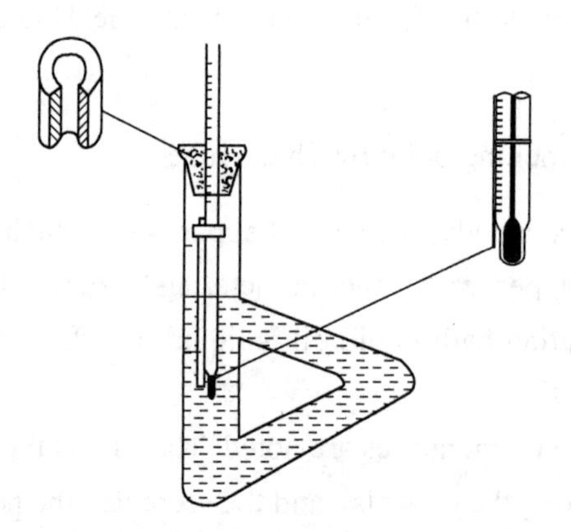

Figure 3.3　Thiele tube device for measuring the melting point

2. Measurement of melting point by microscopic melting point meter

Various types of devices are available on the market for measuring the melting point. The microscopic melting point meters are of particular value when the melting point of a very small amount（e.g. of a single crystal）has to be determined. Further advantages include the possibility of observation of any change in crystalline form of the crystals before melting and the small amount of samples（2-3 crystals）. The main features of the apparatus are shown in Figure 3.4. The rate of heating is controlled by means of a rheostat, and the temperature may be measured covering the ranges 20℃ to 300℃ and with a platinum resistance thermometer and a digital display unit. Before using this melting point apparatus，we must read the guide carefully and strictly follow the operating rules.

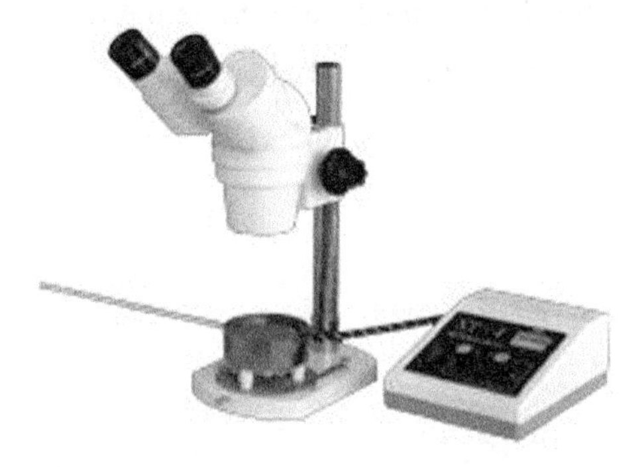

Figure 3.4　Microscopic melting point meter

3. Thermometer correction

The temperatures obtained from the readings of most laboratory thermometers contain errors. For this reason，it is necessary to correct thermometers if temperatures are to be measured accurately. Correction is done by measuring the melting point of several substances and comparing their correct values against the thermometer readings.

【Apparatus and Materials】

Apparatus：Thiele tube（b-type tube），capillary tube，thermometer（≥300℃），rubber stopper（notched），glass tube（70-80 cm），watch glass，rubber band，melting point apparatus.

Materials：liquid paraffin，naphthalene，benzoic acid，carbamide，salicylic acid，*p*-dihydroxybenzene.

【Procedure】

Select one or two compounds from a list of supplied compounds with known melting points. Determine the melting points（ranges）for each of these substances using the Thiele tube apparatus，one time for rough measuring and twice for precise measuring to each compound. Compare and analyze your results with the known melting points.

Select two samples measured from above and mix them thoroughly in 19∶1 or 9∶1 ratio. Determine the melting range of the mixed sample in order to verify the anticipated effects of impurities on the melting range of a pure substance.

Accurately determine and report the melting range of an unknown sample supplied by your instructor. If permitted，determine the same sample by Thiele tube apparatus and microscopic melting point meter respectively for a comparison between the two methods.

3.18　液体有机物折光率测定

【实验目的】

（1）学习测定折光率对研究有机化合物的意义。
（2）掌握使用阿贝折光仪测定液体化合物折光率的方法。

【实验原理】

1. 折光率

折光率是有机化合物的重要物理常数之一。固体、液体和气体都有折光率，尤其是液体有机化合物，文献记载更为普遍。通过测定折光率，可以判断有机化合物的纯度，也可以鉴定未知物。

折光率（n）与物质结构、入射光线的波长、温度、压力等因素有关。通常大气压的变化影响不明显，只是在精密测定时才考虑。使用单色光要比用白光测得更为精准，因此，常用钠光（D 线，$\lambda = 589.3$ nm 或 $\lambda = 589.6$ nm）作光源。折光率的表示需要注明所用光线波长和测定的温度，常用 n_D^{20} 表示，即以钠光为光源，20℃时所测定的 n 值。

通常温度升高（或降低）1℃时，液态有机化合物的折光率就减少（或增加）3.5×10^{-4}～5.5×10^{-4}（通常取均值 4.5×10^{-4}），在实际工作中，常采用粗略的换算公式把某温度下所测得的折光率换算成另一温度下的折光率。

2. 阿贝折光仪

根据不同的需求，市售的折射仪有手持式折光仪、糖量折光仪、蜂蜜折光仪、宝石折光仪、数显折光仪、全自动折光仪及在线折光仪等。在有机化学实验室中，一般用阿贝折光仪来测定折光率，如图 3.5 所示。其工作原理是基于光的折射现象，通过目视望远镜部件和色散校正部件组成的观察部件来调节并找到明暗分界线视场，即临界角的位置，并由数据处理系统将所测得角度转换成折光率显示出来。

【实验仪器与材料】

仪器：阿贝折光仪，恒温水浴锅，擦镜纸。
材料：无水丙酮，无水乙醇，乙酸乙酯，松节油。

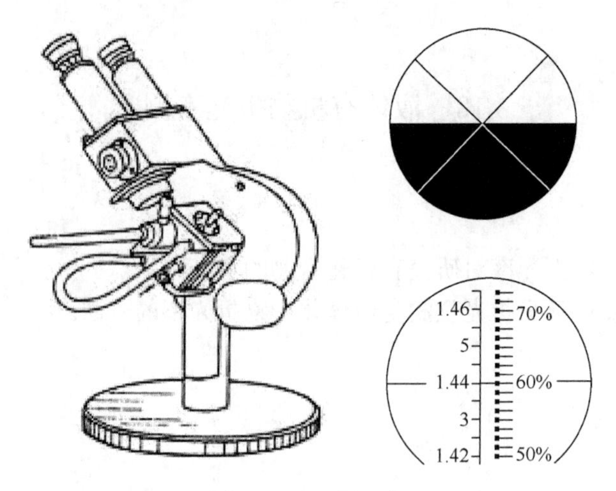

图 3.5　阿贝折光仪

【实验步骤】

开启恒温水浴锅使温度恒定至 20℃。当恒温后，松开锁钮，开启棱镜，滴入 1～2 滴丙酮于镜面上，用擦镜纸蘸少许丙酮轻轻擦拭上、下两棱镜镜面。待完全干燥后，在折射棱镜的抛光面上滴 1～2 滴高纯度蒸馏水，盖上进光棱镜。调节反光镜，使镜筒视场最亮。通过目镜观察视场，同时旋转调节手轮和色散校正手轮，使在目镜中观察到明、暗两部分具有良好的反差和明暗分界线具有最小的色散，视场内明暗分界线准确对准交叉线的交点，如图 3.5 所示。如有偏差，则可用钟表螺丝刀通过色散校正手轮中的小孔，小心旋转里面的螺钉，使分界线相位上下移动至交叉线的交点，然后进行测量，直到读出的纯水的折光率（1.3330）符合标准为止。

打开棱镜，测定无水乙醇、乙酸乙酯和松节油的折光率。当视场内明暗分界线准确对准交叉线的交点时，记录从镜筒中读取的折光率，同时记下温度。重复测定两次，取其平均值为样品的折光率。

仪器用毕后，用沾有少量丙酮的擦镜纸擦拭干净上、下镜面，晾干后合紧两面，用仪器罩盖好，将废弃的擦镜纸收进垃圾桶。

3.18　Determining the Refractive Index of Organic Liquids

【Objectives】

（1）To learn the significance of refractive index in research of organic liquids.

（2）To master the determination of refractive index by using an Abbe refractometer.

【Principles】

1. Definition of refractive index

Refractive index is an important physical constant of organic compounds. The compounds in all possible states—gas，liquid or solid—have their specific refractive index. The determination of refractive index is especially useful and simple for identifying liquids or indicating their purity.

The refractive index，*n*，represents the ratio of the velocity of light in a vacuum（or in air）to the velocity of light in the liquid being studied. The variables of temperature，pressure and the wavelength of the light being refracted influence the refractive index for any substance. Therefore，the temperature（20℃）at which the refractive index was determined is always specified by a superscript in the notation of *n*. The wavelength of light used also affects the refractive index because light of differing wavelengths refracts at different angles. The two bright yellow，closely spaced lines of the sodium spectrum at 589.3 and 589.6 nm，commonly called the sodium D line，usually serves as the standard wavelength for refractive index measurements and are indicated by the subscript D on the symbol *n*. Under these conditions，the refractive index is reported in the form n_D^{20}. If light of some other wavelength is used，the specific wavelength in nanometers appears as the subscript.

Refractive index of liquid always decreases as temperature increased. Variations due to change in temperature are somewhat dependent on the class of compound observed，but are usually somewhere between 3.5×10^{-4}-5.5×10^{-4} per℃. The average value of 4.5×10^{-4} serves as a fair approximation for most liquids.

2. Abbe refractometer

There are different refractometer models，but the Abbe refractometer is one of the most widely used types. A common type of Abbe refractometer in organic lab is shown in Figure 3.5. Its general working principle is based on the above-mentioned phenomenon of the refraction of light when passing through from air to liquid. Refractive index can be read conveniently from the eyepieces after being adjusted correctly.

【Apparatus and Materials】

Apparatus：Abbe refractometer，thermostatic water bath，lens paper.

Materials：anhydrous acetone，anhydrous ethanol，ethyl acetate，turpentine oil.

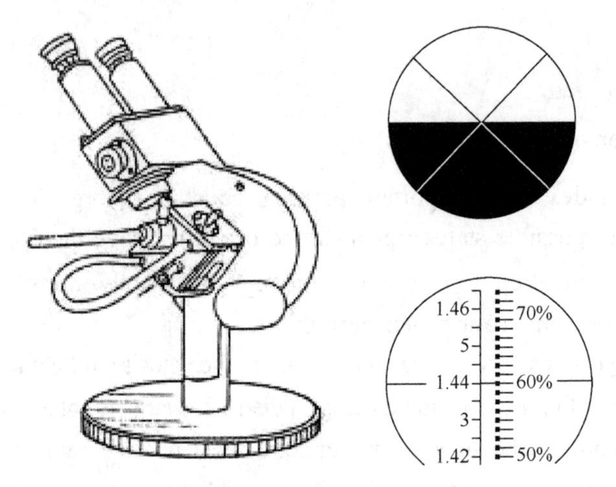

Figure 3.5　Abbe refractometer

【 **Procedure** 】

Begin well the circulation of water from the thermostatic water bath to get a constant temperature（20℃）in advance（If not connected，please skip this step）. Swing open the upper prism gently and check the surface of the prism for cleaning. Place 1-2 drops of acetone on the surface of prism and wipe the upper and lower surfaces with lens paper. When the surfaces are dry，drop 1-2 drops of distilled water onto the lower prism. Lower the upper prism and lock it into position. Look into the eyepieces. Rotate the reflector so it shines through the prism into the sample area. Now adjust the light and the chromic adjustment knob until the field seen in the eyepieces is illuminated so that the light and dark regions are separated by as sharp a boundary as possible（see Figure 3.5）. If the boundary has colors associated with it and/or appears somewhat diffuse，rotate the compensator drum on the face of the instrument until the boundary becomes achromatic and sharp.

Read the value that appears on the field. Reported n_{D}^{20} of distilled water is 1.3330. If there exists an error for the reading，then slowly rotate the screw of the knob from a hole inside until the reading is calibrated to the standard.

Now repeat the above procedures. Measure the refractive index of ethanol，ethyl acetate and turpentine oil respectively. Take twice replicate readings and report the average value.

Taking care not to scratch the surfaces，clean the refractometer prism faces with a soft lens paper moistened with acetone immediately after use. Cover the refractometer before leaving.

3.19　液体有机物比旋光度测定

【实验目的】

（1）掌握使用旋光仪测定手性物质旋光度的方法。

（2）学习比旋光度的计算方法。

【实验原理】

手性化合物使平面偏振光的振动平面向右（“+”）或者向左（“–”）旋转的性质称为旋光性，发生偏转的角度称为旋光度（α）。影响旋光度的因素包括旋光物质的浓度、样品管的长度、溶剂、温度及光源波长。为了比较不同旋光物质的旋光性能，需要有一个统一的比较标准，这个标准就是比旋光度$[\alpha]_{D}^{t}$，它是指在一定测定温度、光源和溶剂条件下的单位旋光度。比旋光度是手性物质特有的物理常数之一。测定比旋光度可以鉴定旋光物质的纯度和含量。

测定旋光度的仪器称为旋光仪，它主要由光源（钠灯）、起偏镜、样品管、检偏镜、目镜和游标刻度盘等部件组成。一般实验室使用的是目测旋光仪，其仪器外形及基本构造如图 3.6 所示。

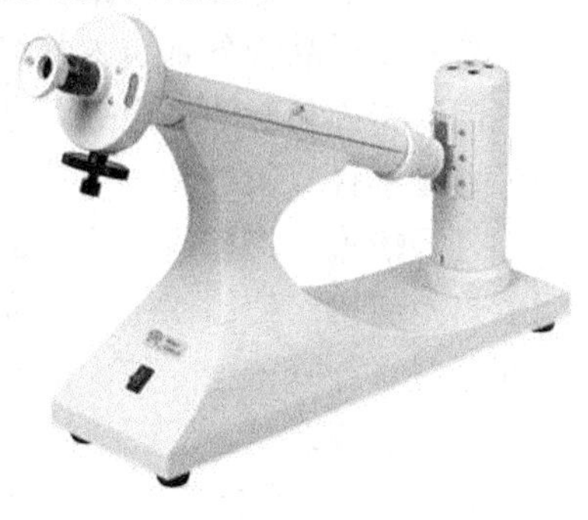

图 3.6　旋光仪

测量时，可从目镜中看到一个明暗相间的三分视场，如图 3.7 所示。通过旋转刻度盘来调整，使视场的中间部分与其两边部分没有明显的界线，强度均匀且整个视场较暗［图 3.7（c）］，判断的标准是稍微来回旋转检偏镜时可观察到三分视场（a）和（b）来回转变，再把图像调回到中间无明显明暗界线的较暗的均一视场，记下分析仪刻度盘读数，记下整数值，再利用游标尺与主盘上刻度线重合的位置，读出游标尺上的数值（小数），可以准确到两位小数。读数示意图见图 3.8。为提高准确度可重复测定几次，取平均值计算其比旋光度或浓度。

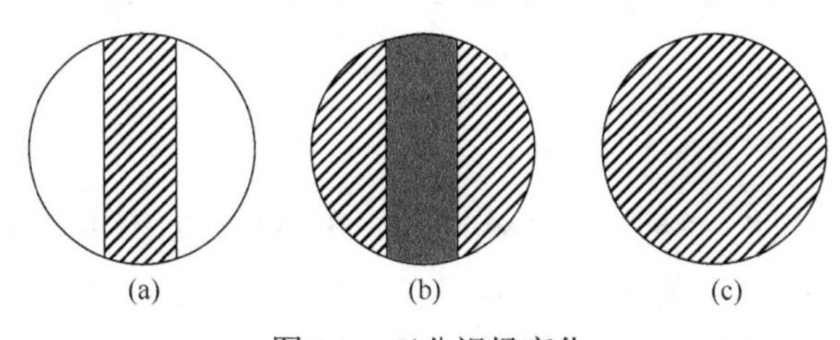

图 3.7　三分视场变化

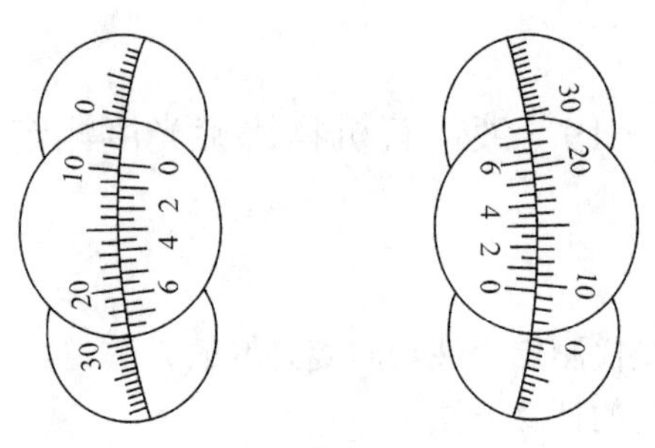

图 3.8　读数示意图

【实验仪器与材料】

仪器：旋光仪，样品管，擦镜纸。

材料：葡萄糖，果糖。

【实验步骤】

打开仪器电源开关，钠光灯亮，预热 5 min。

将三支样品管分别装入蒸馏水、葡萄糖溶液和果糖溶液。样品管中若有气泡，使气泡浮在样品管的突出处，用软布拭净样品管两端，使其有较好的透光性。样品管螺帽不宜旋得过紧。

打开镜盖，将盛有蒸馏水的样品管放入镜筒中，盖上镜盖。观察目镜，同时调节焦距旋钮，使镜筒中三分视场画面锐利清晰。旋转刻度盘手轮直至三分视场各部分明暗程度完全一致。准确读取并记录刻度盘读数，该值为零点校正读数，用于消除系统误差。

再分别将装有葡萄糖溶液和果糖溶液的样品管置于镜筒中同上测定，准确读取刻度盘读数，结合蒸馏水的零点校正读数，确定葡萄糖溶液和果糖溶液的比旋光度。

根据所测比旋光度值计算葡萄糖和果糖溶液的浓度（注：校准读数应排除在外！）。

3.19　Determining the Specific Rotation of Organic Liquids

【**Objectives**】

（1）To master the measurement of the optical rotation of the chiral compounds by using the polarimeter.

（2）To learn the meaning of specific rotation.

【 **Principles** 】

Chiral compounds distinguish themselves with other compounds by their ability of changing the oscillating direction of a beam of plane-polarized light. They can rotate the oscillating plane to the right(labeled as "+")or to the left("−"), and the extent of the rotation is measured as rotation angle (α) . The rotation angle for one compound is influenced by a series of factors: concentration of the rotatory substance, length of sample tube, solvent, temperature and wavelength of the light source. All these factors should be taken into account in the measurement. In order to estimate the comparable potency of optical rotation between different chiral compounds, all the influencing factors should be placed at a standard level in measuring, and the rotation measured under such standard conditions is called specific rotation, labeled as $[\alpha]_{\mathrm{D}}^{t}$. Specific rotation is a characteristic physical constant of the chiral substances which is recorded by most of the handbooks. The determination of rotation angle can be used to identify the optical purity of chiral compounds or calculate their contents in the solution.

The instrument used for the measurement of rotation angle is called polarimeter, which is composed of five main working parts: light source (sodium lamp), polarizer, sample tube, analyzer, eyepiece and vernier scale (see Figure 3.6) .

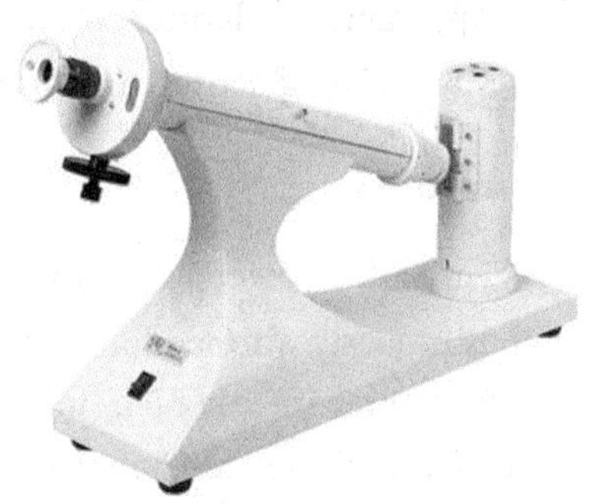

Figure 3.6　Polarimeter

A three-part field of view between light and dark in the eyepiece can be seen, as shown in Figure 3.7.

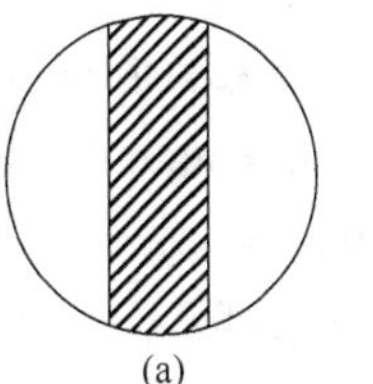

(a)

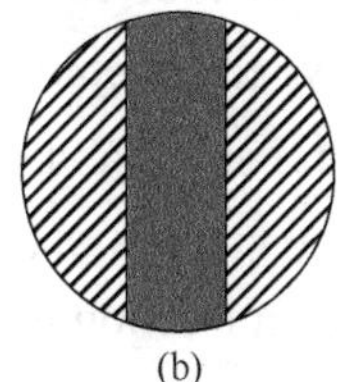

(b)

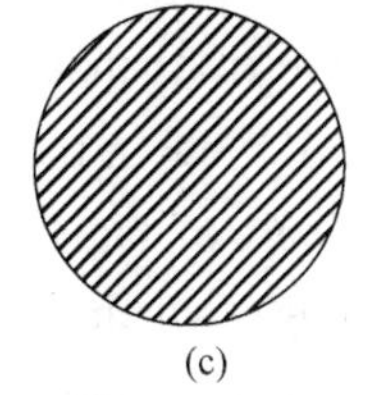

(c)

Figure 3.7　Split-fields and uniform field for reading

Adjust to sharpen the focus first, then rotate the handwheel clockwise (to the right) or counterclockwise (to the left) until two fields are clearly visible (often a vertical bar split down the middle of a background field) and interchange quickly in a small rotation range. Then back off a little slowly until no obvious split-image and the entire visual field is as uniform as possible [see Figure 3.7（c）]. Read the rotation angle from the analyzer scale and use the vernier scale to estimate the reading to a fraction of a degree（see Figure 3.8）. Repeat the determination several times and use the average value to calculate its specific rotation or concentration.

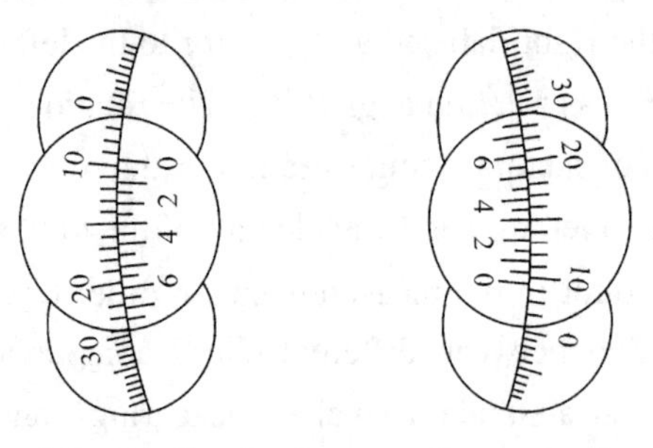

Figure 3.8　Reading the demonstration on vernier scale

【Apparatus and Materials】

Apparatus：polarimeter，sample tubes，lens paper.

Materials：glucose，fructose.

【Procedure】

Throw the power switch of the polarimeter to the “ON” position until the sodium lamp is properly warmed for 5 min.

Pack three sample tubes into distilled water, glucose solution and fructose solution respectively. If there are bubbles in the sample tube，the bubbles float in the protrusion of the sample tube. Wipe the ends of the sample tube with soft cloth to make it have better transparency. The nut of pack sample tube should not be screwed too tightly.

Open the mirror cap, put the sample tube containing distilled water into the mirror tube, and cover the mirror cap. Observe the eyepiece and adjust the focus knob to make the three-part field of view in the barrel sharp and clear. Rotate the dial wheel until the brightness of each part of the field of view is exactly the same. Accurately read and record the dial readings, which are zero correction readings for eliminating system errors.

The sample tubes containing glucose solution and fructose solution were placed in the mirror tube respectively. The dial reading was read accurately，and the zero correction reading of distilled water was used to determine the specific rotation of the glucose solution and fructose solution.

Calculate the concentration of glucose or fructose solution according to the measured rotation angle（note the calibrating reading should be excluded!）and their specific rotation.

3.20　有机化合物官能团鉴定实验

【实验目的】

（1）理解有机化合物主要官能团的特征化学性质。
（2）学习典型有机化合物的定性鉴别分析方法。

【实验原理】

有机化合物分子中的官能团是分子中比较活泼而容易发生化学反应的部位。通过官能团所特有的化学现象，能够验证和区别官能团的种类。有机化合物各种官能团能发生的化学反应很多，但能用于有机化学分析中的特征反应应具备以下条件：①反应迅速；②反应的性质变化易于观察，如颜色、溶解、沉淀、气体逸出等；③灵敏度高；④专一性强。

【实验仪器与材料】

仪器：烧杯，试管，试管架，滴管，水浴锅，显微镜，电磁搅拌加热板。
材料：

分类	材料
卤代烃的鉴定	C_6H_5Cl，$CH_3CH_2CH_2Cl$，$CH_2=CHCH_2Cl$，C_2H_5OH，$AgNO_3$
醇、酚的鉴定	C_2H_5OH，C_6H_5OH，$CuSO_4$，NaOH，$FeCl_3$，β-萘酚，甘油，邻苯二酚
醛、酮的鉴定	HCHO，CH_3CHO，CH_3COCH_3，C_6H_5CHO，$(CH_3)_2CHOH$，I_2，NaOH，费林试剂（Ⅰ），费林试剂（Ⅱ），2, 4-二硝基苯肼
酸类的鉴定	HCOOH，CH_3COOH，HOOC—COOH，Na_2CO_3，NaOH，HCl，$FeCl_3$，饱和溴水，苯甲酸，水杨酸，乙酰乙酸乙酯
胺类的鉴定	CH_3NH_2，$C_6H_5NH_2$，HCl，浓 HCl，酚酞指示剂
糖类的鉴定	$AgNO_3$，NaOH，HCl，苯肼，浓 $NH_3 \cdot H_2O$，浓 HCl，浓 H_2SO_4，C_2H_5OH，葡萄糖，麦芽糖，果糖，蔗糖，淀粉，α-萘酚，乙酸钠，费林试剂（Ⅰ），费林试剂（Ⅱ）
氨基酸、蛋白质的鉴定	NaOH，HAc，$(NH_4)_2SO_4$，$CuSO_4$，$AgNO_3$，浓 HCl，甘氨酸，酚酞指示剂，甲基橙指示剂，蛋清蛋白液，茚三酮

【实验步骤】

1. 卤代烃的鉴定

取 3 支干燥试管并做好标记，分别加入 2 滴 20% C_6H_5Cl，20% $CH_3CH_2CH_2Cl$，20% $CH_2=CHCH_2Cl$，再加入 2～4 滴饱和 $AgNO_3$-C_2H_5OH 溶液，充分摇匀，观

察有无沉淀生成。将无沉淀生成的试管放置于水浴（70℃）中加热 3～5 min，再观察是否有沉淀生成。根据实验现象，归纳不同结构卤代烃与 $AgNO_3$ 的反应次序，并解释原因。

2. 醇、酚的鉴定

（1）多元醇与 $Cu(OH)_2$ 的作用。取 2 支试管，分别加入 5 滴 5% $CuSO_4$ 和 5 滴 10% NaOH 溶液，摇匀后，在其中一支试管中加入 1 mL 95% C_2H_5OH，在另一支试管中加入 1 mL 甘油，摇匀。观察现象并解释原因。

（2）酚与 $FeCl_3$ 的显色反应。取 3 支试管，分别加入 3 滴 5% C_6H_5OH，5%邻苯二酚，5% β-萘酚，然后各加入 2 滴 5% $FeCl_3$ 溶液，观察其颜色变化情况。

3. 醛、酮的鉴定

（1）与 2, 4-二硝基苯肼反应。取 4 支试管，各加入 5 滴 2, 4-二硝基苯肼，然后分别加入 2 滴 HCHO、CH_3CHO、CH_3COCH_3、C_6H_5CHO 溶液，微微振荡，观察是否有沉淀产生。如无沉淀出现，可在温水浴中加热 30 s，冷却后观察变化情况。

（2）与费林试剂反应。取 4 支试管，各加入 5 滴费林试剂（Ⅰ）和 5 滴费林试剂（Ⅱ），摇匀，得深蓝色透明液体。然后分别加入 10 滴 HCHO、CH_3CHO、CH_3COCH_3、C_6H_5CHO 溶液，摇匀，放置于沸水浴中加热 3～5 min。观察溶液颜色有无变化，有无沉淀产生。

（3）碘仿反应。取 4 支试管，分别加入 10 滴 95% C_2H_5OH、CH_3COCH_3、$(CH_3)_2CHOH$、CH_3CHO，再各加入 10 滴碘液，然后边振荡边逐滴加入 5% NaOH 溶液至棕色刚好褪去，观察是否有黄色沉淀生成。若无沉淀生成，置于水浴中微热后，再观察有无黄色沉淀生成。根据实验现象，归纳出能发生碘仿反应的化合物的结构特点。

4. 羧酸和羧酸衍生物的鉴定

（1）在 3 支试管中各加入 2 mL 10% Na_2CO_3 溶液，再分别加入 5 滴 10%甲酸、10%乙酸、10%草酸，摇匀，观察是否有气体逸出现象，并解释。

（2）在 1 支试管中加入 0.1 g 固体苯甲酸，加 1 mL 蒸馏水，振摇，观察固体是否溶解。再加入数滴 10% NaOH 溶液，振摇，观察现象。然后加入 10%盐酸数滴，观察又有何现象，说明原因。

（3）与 $FeCl_3$ 反应。取 2 支试管，分别加入 5 滴饱和苯甲酸溶液、5 滴饱和水杨酸溶液，再各加 2 滴 1% $FeCl_3$ 溶液，摇匀。观察各试管内有何现象，说明原因。

（4）酮-烯醇式互变。取 1 支试管，加入 1 mL 蒸馏水，5 滴乙酰乙酸乙酯溶液，振荡。再加入 2 滴 5% $FeCl_3$，摇匀，观察其颜色变化。然后滴加饱和溴水（用量不可太多），摇匀，可观察到紫红色褪去，放置一会儿，再观察其颜色是否重现，解释原因。

5. 胺、酰胺的鉴定

（1）取 1 支试管，加入 3 滴 CH_3NH_2 溶液，再加入 1 滴 0.1%酚酞指示剂，摇匀。观察其现象。然后逐滴加入 5% HCl 溶液，又有何变化？

（2）取 1 支试管，加入 3 滴 $C_6H_5NH_2$ 再加入 1 mL 蒸馏水，此时 $C_6H_5NH_2$ 溶解吗？然后边振荡边逐滴加入浓 HCl，观察其现象，解释原因。

6. 糖的鉴定

（1）α-萘酚试验（Molish 试验）。在 5 支试管中分别加入 0.5 mL 4%葡萄糖、4%果糖、4%麦芽糖、4%蔗糖和 2%淀粉溶液，各滴入 2 滴 α-萘酚-乙醇溶液，摇匀，倾斜试管 45°，沿管壁慢慢加入 1 mL 浓 H_2SO_4，勿摇动，H_2SO_4 在下层，样品在上层，观察两层交界处有何现象，解释原因。

（2）银镜反应。取 5 支洁净的试管，各加入 1 mL 5% $AgNO_3$ 溶液和 2 滴 20% NaOH 溶液，逐滴加入浓 $NH_3 \cdot H_2O$ 至生成的沉淀恰好溶解，摇匀。再分别加入 0.5 mL 4%葡萄糖、4%果糖、4%麦芽糖、4%蔗糖和 2%淀粉溶液，摇匀后，在 60℃水浴中加热 10～15min，观察有无银镜生成，根据实验现象，归纳特点。

（3）成脎反应。在 4 支试管中分别加入 1 mL 4%葡萄糖、4%果糖、4%麦芽糖、4%蔗糖溶液，再各加入 0.5 mL 盐酸苯肼-乙酸钠试剂，摇匀后放入沸水浴中加热 20～30 min，冷却比较成脎结晶的速度和颜色，并在显微镜下观察糖脎的结晶形状。

（4）淀粉水解反应。在 1 支试管中加入 3 mL 2%淀粉溶液，再加入 5 滴浓 HCl 溶液，摇匀。于沸水浴中加热 5～8 min，冷却后用 10% NaOH 中和，加费林试剂（Ⅰ）和费林试剂（Ⅱ）各 2 滴，沸水浴加热后观察现象，解释原因。

7. 氨基酸、蛋白质的鉴定

（1）酸碱两性。在 2 支试管中各加入 3 mL 蒸馏水，一支试管中加入 2 滴 10% NaOH、1 滴酚酞指示剂，另一支试管中加入 2 滴 10% HAc 溶液和 1 滴甲基橙指示剂，然后分别加入 1 mL 2%甘氨酸溶液，摇匀。观察颜色变化并解释原因。

（2）盐析作用。在 1 支试管中加入 3 mL 蛋清蛋白质溶液，再加入 0.5 g $(NH_4)_2SO_4$ 晶体使其成为 $(NH_4)_2SO_4$ 的饱和溶液，观察现象。再加入 1 mL 蒸馏水，振荡，观察现象并解释原因。

（3）蛋白质变性。在 3 支试管中各加入 2 mL 蛋清蛋白液，再分别加 5 滴浓盐酸、5% $CuSO_4$、2% $AgNO_3$ 溶液，摇匀，观察沉淀的生成。再各加 1 mL 蒸馏水，观察沉淀是否溶解。

（4）茚三酮反应。在 2 支试管中分别加入 1 mL 2%甘氨酸和 1 mL 蛋清蛋白液，然后分别加入 10 滴 1%茚三酮溶液，摇匀，然后将试管放入沸水浴中加热 10～15 min，观察现象并解释原因。

8. 有机定性分析——辨别未知物

有 A、B、C、D、E 五瓶液体，已知它们是甲醇、乙醇、正丙醇、乙醛或丙酮，但不知各瓶中装的是哪一种。请设计一个分析方案并用实验证实 A～E 瓶中各装的是什么物质。

3.20 Identification of Organic Compounds

【Objectives】

（1）To understand the reaction characters of the main functional groups of the major organic compounds.

（2）To learn how to identify the class of compounds which the unknown belong to.

【Principles】

Functional groups in organic compounds are molecules that are more reactive and prone to chemical reactions. We can verify the properties of various functional groups through their peculiar chemical phenomena. Organic compounds have a lot of chemical reactions of various functional groups，but the reactions used in organic chemical analysis should have the following conditions：①rapid reaction；②the reaction should be easily observed in its properties，such as color，dissolution，precipitation，gas escaping；③high sensitivity；④high specificity（reagents and functional groups reaction）.

【Apparatus and Materials】

Apparatus：beaker，test tubes，test tube holder，dropper，water bath，microscope，magnetic stirring electric heating plate.

Materials：

Classification	Materials
identification of halohydrocarbon	C_6H_5Cl, $CH_3CH_2CH_2Cl$, $CH_2=CHCH_2Cl$, C_2H_5OH, $AgNO_3$
identification of alcohol, phenol	C_2H_5OH, C_6H_5OH, $CuSO_4$, NaOH, $FeCl_3$, *β*-naphthol, glycero, *o*-dihydroxybenzene
identification of aldehyde, ketone	HCHO, CH_3CHO, CH_3COCH_3, C_6H_5CHO, $(CH_3)_2CHOH$, I_2, Fehling reagent（Ⅰ）, Fehling reagent（Ⅱ）, 2, 4-dinitrophenylhydrazine
identification of carboxylic acid	HCOOH, CH_3COOH, HOOC—COOH, Na_2CO_3, NaOH, HCl, $FeCl_3$, Br_2, benzoic acid, salicylic acid, acetyl ethyl acetate
identification of amine, acid amide	CH_3NH_2, $C_6H_5NH_2$, HCl, concentrated HCl, phenolphthalein reagent
identification of sugar	$AgNO_3$, NaOH, HCl, $C_6H_5NHNH_2$, concentrated $NH_3 \cdot H_2O$, concentrated HCl, concentrated sulfuric acid, ethanol, glucose, fructose, maltose, sucrose, starch, *α*-naphthol, sodium acetate, Fehling reagent（Ⅰ）, Fehling reagent（Ⅱ）
identification of amino acid, protein	NaOH, HAc, $(NH_4)_2SO_4$, $CuSO_4$, $AgNO_3$, concentrated HCl, glycine, phenolphthalein reagent, methyl orange reagent, egg protein, ninhydrin

【Procedure】

1. Identification of halohydrocarbon

Take three dry test tubes and label them in sequence. Add 2 drops of 20% C_6H_5Cl, 20% $CH_3CH_2CH_2Cl$, 20% $CH_2=CHCH_2Cl$ in the test tube respectively, and then add 2-4 drops of saturated ethanol solution of $AgNO_3$ in each test tube. Observe if there are precipitates appearing in sequence during shaking, otherwise, place the test tubes in warm water（70℃）for 3-5 min. Observe the formation of precipitates. According to the results, rank the relativity of the above three halohydrocarbons in the reaction with $AgNO_3$ and give an explanation.

2. Identification of alcohols and phenols

（1）Reaction of polyhydric alcohol and copper hydroxide. Take two test tubes. Add 5 drops of 5% $CuSO_4$ solution respectively, and then add about 5 drops of 10% NaOH in each tube by shaking. Then add 1 mL 95% C_2H_5OH into the first test tube, 1 mL glycerol into the second test tube. Shake the mixture and then let them stand for a while. Observe the phenomenon of the mixture respectively and give an explanation.

（2）Ferric chloride test for phenols. Take three test tubes. Add 3 drops of 5% C_6H_5OH, 5% *o*-dihydroxybenzene, 5% *β*-naphthol respectively, and then place 2 drops of 5% $FeCl_3$ in each tube. Observe the changes shown in every tube.

3. Identification of aldehydes and ketones

（1）Phenylhydrazine reactions. Take four test tubes. Add 5 drops of 2, 4-

dinitrophenylhydrazine in each test tubes，and then add 2 drops of HCHO，CH_3CHO，CH_3COCH_3，C_6H_5CHO solution in each tube respectively. Shake them slightly and then stand for a while. Observe if the precipitate is formed in each tube. However，if the precipitate does not form immediately, let the solution stand for 30 s in warm water. And then cool it in the room temperature after shaking it. Observe the change.

（2）Fehling tests. Take four test tubes. Add 5 drops of Fehling reagent(Ⅰ) and 5 drops of Fehling reagent（Ⅱ）in each tube and shake them respectively. Obtain the dark blue transparent liquid. And then add 10 drops of HCHO, CH_3CHO, CH_3COCH_3, C_6H_5CHO solution with shaking，respectively. Heat them in the boiling water bath for 3-5 min. Observe the color or the precipitate of change.

（3）Iodoform tests. Take four test tubes. Add 10 drops of 95% C_2H_5OH，CH_3COCH_3，$(CH_3)_2CHOH$，CH_3CHO solution respectively，and then add 10 drops of iodine reagent in each tube. And then add 5% NaOH in each tube drop by drop with shaking until the brown color of the solution disappears. However，if a precipitate does not form immediately，let the solution stand in warm water. Observe whether the yellow iodoform is formed. According to the results，rank the relativity of the above iodoform test.

4. Identification of carboxylic acids and their derivatives

（1）Take three test tubes. Add 2 mL 10% Na_2CO_3 solution in the each test tube，then add 5 drops of 10% formic acid solution，10% acetic acid solution，10% oxalic acid solution into the tubes with shaking，respectively. Observe the change and give an explanation.

（2）Take one test tube. Add 0.1 g benzoic acid（solid）and 1 mL water in the test tube with shaking. Observe the change. Then cautiously add dropwise 10% NaOH solution with shaking. Observe the change. Then add several drops of 10% HCl into the tube to see what will happen. Explain all the changes.

（3）Reaction of acid with $FeCl_3$. Take two test tubes. Add 5 drops of saturated benzoic acid solution and 5 drops of saturated salicylic acid solution respectively，then add 2 drops of 1% $FeCl_3$ with shaking. Observe and explain the color change of the solution.

（4）Keto-enol tautomerism of ethyl acetoacetate. Take one test tube. Add 1 mL water and 5 drops of acetyl ethyl acetate in the test tube with shaking. Then add 2 drops of 5% $FeCl_3$ with shaking. Observe the color of change. Then add saturated bromine water，what phenomenon do you observe? And what change has taken place after waiting a while? Please explain the series of changes.

5. Identification of amines and acid amides

（1）Take one test tube. Add 3 drops of CH_3NH_2 and 1 drop of 0.1% phenolphthalein solution in the test tube with shaking. Observe the color of change. Then add 5% HCl solution. Please explain the series of changes.

（2）Take one test tube. Add 3 drops of $C_6H_5NH_2$ solution and 1 mL water in the test tube with shaking. Observe the change of solution. Then cautiously add dropwise concentrated HCl solution. Observe the series of experimental phenomena，compare and explain the results.

6. Identification of sugar

（1）α-naphthol test（Molish test）. Take five test tubes. 0.5 mL 4% glucose，4% fructose，4% maltose，4% sucrose and 2% starch solution are added into the test tubes respectively，and then add 2 drops of α-naphthol-ethanol solution with shaking. The test tube is inclined 45° after mixing and 1 mL concentrated sulfuric acid（Do not shake!）is added along the wall slowly. Sulfuric acid is at the bottom and the test tube solution is at the up layer. Observe the change between the layers. Explain the results.

（2）Silver mirror reaction. Take five test tubes. Add 1 mL 5% $AgNO_3$ and 2 drops of 20% NaOH solution respectively，then cautiously add dropwise concentrated $NH_3 \cdot H_2O$ with shaking. Continue adding until the precipitates disappear completely in each tube. Add and mix 0.5 mL 4% glucose，4% fructose，4% maltose，4% sucrose and 2% starch solution in the test tubes respectively. Heat them in the water bath（60℃）for 10-15 min. Observe the experimental phenomena，compare and explain the results.

（3）Osazone reaction. Take four test tubes. Add 1 mL 4% glucose，4% fructose，4% maltose，4% sucrose solution，respectively，then add 0.5 mL of mixture solution of phenylhydrazine hydrochloride and sodium acetate in each tube. Shake the tubes to fully mix the solution，then heat them in a boiling water bath for 20-30 min. Take them out and cool for a while to allow the reaction products to crystalize from the solutions. Observe the formations of osazone of different saccharides. Compare the crystalline rate and color between 4 kinds of osazone. And drop a little of osazone of each on glass plates to observe their crystal forms under a microscope.

（4）Starch hydrolysis reaction. Take one test tube. Add 3 mL 2% starch solution and 5 drops of concentrated HCl with shaking. Heat them in the boiling water bath for 5-8 min. And then cool it in the room temperature and neutralize by adding 10% NaOH solution. Mix 2 drops of Fehling reagent(Ⅰ) and 2 drops of Fehling reagent（Ⅱ）together and then heating them in a boiling water bath. Observe the experimental phenomena and explain the results.

7. Identification of amino acids and proteins

（1）Amphoteric molecule. Take two test tubes. Add 3 mL water in the each test tube. Then add 2 drops of 10% NaOH solution and 1 drops of phenolphthalein solution in the first test tube with shaking. Then add 2 drops of 10% HAc solution and 1 drop of methyl orange solution in the second test tube with shaking. Mix 1 mL 2% glycine solution into each test tube with shaking respectively. Observe the series of experimental phenomena，compare and explain the results.

（2）Salting-out effect. Take one test tube. Add 3 mL egg protein solution. Then add 0.5 g$(NH_4)_2SO_4$ into the test tube. Then add 1 mL water with shaking. Observe the series of experimental phenomena，compare and explain the results.

（3）Denaturation of proteins. Take three test tubes. Add 2 mL egg protein solution respectively，and then add 5 drops of concentrated HCl，5% $CuSO_4$，2% $AgNO_3$ solution carefully drop by drop with shaking respectively. Observe what happened in the test tubes. Add 1 mL water again. Observe the results and explain.

（4）Ninhydrin reaction. Take two test tubes. Add 1mL of 2% glycine and 1 mL of egg protein solution respectively，and then add 10 drops of 0.1% ninhydrin solution into each test tube with shaking. Heat them in the boiling water bath for 10-15 min. Observe and compare the results and explain.

8. Identification of unknown compounds

Now，there are five bottles of liquids（labeled A，B，C，D and E）on the table. We just know these liquid compounds are：methanol（CH_3OH），ethanol（CH_3CH_2OH），propanol（$CH_3CH_2CH_2OH$），acetaldehyde（CH_3CHO）and acetone（CH_3COCH_3），but we don't know exactly which bottle contains which compound. Please design a reasonable and simple scheme and carry out some corresponding tests to identify them one by one.

参 考 文 献

References

陈彪，魏永惠. 2013. 有机化学实验（英汉双语教材）. 北京：化学工业出版社.

陈秋云. 2016. Experimental Organic Chemistry. 镇江：江苏大学出版社.

冯文芳. 2014. 有机化学实验（双语）. 武汉：华中科技大学出版社.

高岩，洪波. 2007. 有机化学实验. 北京：中国农业出版社.

郭书好. 2006. 有机化学实验. 2 版. 武汉：华中科技大学出版社.

李雯良. 2003. 微型半微型有机化学实验. 北京：高等教育出版社.

李英俊，孙淑琴. 2009. 半微量有机化学实验（中英文对照）. 2 版. 北京：化学工业出版社.

唐冬雁，刘本才. 2005. 应用化学专业英语. 3 版. 哈尔滨：哈尔滨工业大学出版社.

薛思佳，季萍，Olson L. 2016. 有机化学实验（英汉双语版）. 3 版. 北京：科学出版社.

袁华，尹传奇. 2014. 有机化学实验（双语版）. 北京：化学工业出版社.

张大伟. 2010. 大学综合化学实验指导. 长春：吉林大学出版社.

Bettelheim F A，Landesberg J M. 2000. Laboratory Experiments for General，Organic，and Biochemistry. 4th ed. New York：Harcourt Inc.

Mohrig J R，Hammond C N，Schatz P F. 2010. Techniques in Organic Chemistry. 3rd ed. New York：W. H. Freeman and Company.

Schwetlick K. 2012. 有机合成实验室手册. 22 版. 万均等译. 北京：化学工业出版社.

Wang M，Wang Y H，Gao Z X. 2011. Organic Chemistry Experiments（有机化学实验）. 北京：高等教育出版社.

附 录
Appendix

Ⅰ. 常用元素的相对原子质量

原子序数	元素符号	英文名称	中文名称	原子序数	元素符号	英文名称	中文名称
1	H	hydrogen	氢	19	K	potassium	钾
2	He	helium	氦	20	Ca	calcium	钙
3	Li	lithium	锂	21	Sc	scandium	钪
4	Be	beryllium	铍	22	Ti	titanium	钛
5	B	boron	硼	23	V	vanadium	钒
6	C	carbon	碳	24	Cr	chromium	铬
7	N	nitrogen	氮	25	Mn	manganese	锰
8	O	oxygen	氧	26	Fe	iron	铁
9	F	fluorine	氟	27	Co	cobalt	钴
10	Ne	neon	氖	28	Ni	nickel	镍
11	Na	sodium	钠	29	Cu	copper	铜
12	Mg	magnesium	镁	30	Zn	zinc	锌
13	Al	aluminum	铝	35	Br	bromine	溴
14	Si	silicon	硅	45	Rh	rhodium	铑
15	P	phosphorus	磷	46	Pd	palladium	钯
16	S	sulfur	硫	47	Ag	silver	银
17	Cl	chlorine	氯	53	I	iodine	碘
18	Ar	argon	氩	79	Au	gold	金

Ⅱ. 有机化学实验常用仪器、实验操作

1. 常用普通仪器

烧杯 beaker

试管 test tube

滴管 dropper

移液管，吸液管 suction pipet；pipette

提勒管 Thiele tube

烧瓶 flask

圆底烧瓶 round flask；round-bottom flask

平底烧瓶 florence flask

碘量瓶 iodine flask

蒸馏烧瓶 distilling flask

毛细管 capillary tube
干燥管 drying tube
接液管 distillation adapter；adapter
双排管 double-blank manifold
量杯 measuring cup
量筒 measuring cylinder
漏斗 funnel
玻璃漏斗 glass funnel
分液漏斗 separating funnel
布氏漏斗 Büchner funnel
长颈漏斗 long-stem funnel
滴液漏斗 dropping funnel
广口瓶 wide-mouth bottle
锥形瓶 Erlenmeyer/conical flask
试剂瓶 reagent bottle
抽滤瓶 suction flask
温度计 thermometer
薄层板 thin layer plate
硅胶板 silica gel plate
滤纸 filter paper
橡皮塞 rubber stopper
带缺口的橡皮塞 rubber stopper（notched）
微型注射器 micro-syringe
旋转蒸发器 rotatory evaporator
索氏提取器 Soxhlet extractor
三颈圆底烧瓶 three-neck round flask
蒸馏头 distilling head
真空蒸馏头 vacuum-distilling head
克氏蒸馏头 Claisen distilling head
球形冷凝管 Allihn condenser
直形冷凝管 straight condenser
空气冷凝管 air /liebig condenser
蛇形冷凝管 graham condenser
分馏柱 fractionating column
分馏头 fractionating head
烧瓶夹 flask clamp
铁架台 iron support stand
搅拌子（棒）stir bar
干燥器 desiccator
研钵 mortar
蒸发皿 evaporation dish
表面皿 watch glass
水浴锅 water bath
层析缸 chromatographic tank
层析柱 chromato bar
镊子 tweezers
搅拌器 agitator
磁力搅拌装置 magnetic stirring apparatus
减压抽滤装置 vacuum filtration apparatus
电磁搅拌加热套 magnetic stirring electric heating mantle

电热套 electric heating mantle

电磁搅拌加热板 magnetic stirring electric heating plate

电热板 electric heating plate

回流装置 reflux apparatus

烘箱 oven

蒸馏装置 simple distillation apparatus

真空泵 vacuum pump

萃取装置 extraction apparatus

抽气泵 aspirator pump

熔点仪 melting point apparatus

分析天平 analytical balance

阿贝折光仪 Abbe refractormeter

护目镜 goggles

旋光仪 polarimeter

紫外灯 ultraviolet（UV）lamp

超声波清洗器 ultrasonic cleaner

2. 常用实验操作

升华 sublimation

萃取 extraction

中和 neutralize

干燥 dry

减压蒸馏 vacuum distillation

重结晶 recrystallization

蒸馏 simple distillation

回流 reflux

水蒸气蒸馏 steam distillation

脱水 dehydration

分馏 fractional distillation

水解 hydrolysis

过滤 filtration

机械搅拌 mechanical stirring

减压过滤 vacuum filtration

磁力搅拌 magnetic stirring

3. 常用大型仪器

中文名称	英文名称	英文缩写
原子发射光谱仪	atomic emission spectrometer	AES
原子吸收光谱仪	atomic absorption spectrometer	AAS
原子荧光光谱仪	atomic fluorescence spectrometer	AFS
紫外-可见分光光度计	UV-visible spectrophotometer	UV-Vis
傅里叶变换红外光谱仪	FT-IR spectrometer	FTIR
气相色谱仪	gas chromatograph	GC
高压（效）液相色谱仪	high pressure（performance）liquid chromatograph	HPLC
凝胶渗透色谱仪	gel permeation chromatograph	GPC
质谱仪	mass spectrometer	MS

续表

中文名称	英文名称	英文缩写
气相色谱-质谱联用仪	gas chromatograph mass spectrometer	GC-MS
液相色谱-质谱联用仪	liquid chromatograph mass spectrometer	LC-MS
核磁共振波谱仪	nuclear magnetic resonance spectrometer	NMR
电感耦合等离子体发射光谱仪	inductively coupled plasma emission spectrometer	ICP-OES
电感耦合等离子质谱仪	inductively coupled plasma mass spectrometer	ICP-MS

III. 有机化学实验常用溶剂性质和缩写

中文名称	英文名称	缩写	溶解性	毒性
甲胺	methylamine	MY	多数有机物和无机物的优良溶剂，液态甲胺与水、醚、苯、丙酮、低级醇混溶，其盐酸盐易溶于水，不溶于醇、醚、酮、氯仿、乙酸乙酯	中等毒性，易燃
甲醇	methanol	MeOH	与水、乙醚、醇、酯、卤代烃、苯、酮混溶	中等毒性，麻醉性
二氯甲烷	dichloromethane	DCM	与醇、醚、氯仿、苯、二硫化碳等有机溶剂混溶	低毒，麻醉性强
石油醚	petroleum ether	PE	不溶于水，与丙酮、乙醚、乙酸乙酯、苯、氯仿及甲醇以上高级醇混溶	与低级烷相似
乙醚	diethyl ether	DEE	微溶于水，易溶于盐酸，与醇、醚、石油醚、苯、氯仿等多数有机溶剂混溶	麻醉性
乙醇	ethanol	EtOH	与水、乙醚、氯仿、酯、烃类衍生物等有机溶剂混溶	微毒，麻醉性
乙腈	acetonitrile	ACN	与水、甲醇、乙酸甲酯、乙酸乙酯、丙酮、醚、氯仿、四氯化碳、氯乙烯及各种不饱和烃混溶，但不与饱和烃混溶	中等毒性，蒸气易引起急性中毒
乙酸乙酯	ethyl acetate	EA	与醇、醚、氯仿、丙酮、苯等大多数有机溶剂混溶，能溶解某些金属盐	低毒，麻醉性
丙酮	acetone	DMK	与水、醇、醚、烃混溶	低毒，类乙醇
氯仿	chloroform	TCM	与乙醇、乙醚、石油醚、卤代烃、四氯化碳、二硫化碳等混溶	中等毒性，强麻醉性
四氯化碳	carbon tetrachloride	CT	与醇、醚、石油醚、石油脑、冰醋酸、二硫化碳、氯代烃混溶	氯化甲烷中毒性最强
四氢呋喃	tetrahydrofuran	THF	优良溶剂，与水混溶，与乙醇、乙醚、脂肪烃、芳香烃、氯化烃等混溶	吸入微毒，经口低毒
己烷	hexane	HX	甲醇部分溶解，与比乙醇高的醇、醚丙酮、氯仿混溶	低毒，麻醉性，刺激性
环己烷	cyclohexane	CYH	与乙醇、高级醇、醚、丙酮、烃、氯代烃、高级脂肪酸、胺类混溶	低毒，中枢抑制作用

续表

中文名称	英文名称	缩写	溶解性	毒性
1, 4-二氧六环	1, 4-dioxane	Diox	能与水及多数有机溶剂混溶，溶解能力强	微毒，强于乙醚2～3倍
苯	benzene	Phe	难溶于水，与甘油、乙二醇、乙醇、氯仿、乙醚、四氯化碳、二硫化碳、丙酮、甲苯、二甲苯、冰醋酸、脂肪烃等大多有机物混溶	强烈毒性
甲苯	toluene	PhMe	不溶于水，与甲醇、乙醇、氯仿、丙酮、乙醚、冰醋酸、苯等有机溶剂混溶	低毒，麻醉性
二乙胺	diethylamine	DEA	溶于水、乙醇、苯和乙醚，微溶于庚烷	刺激皮肤、眼睛
N, *N*-二甲基甲酰胺	*N*, *N*-dimethyl-formamide	DMF	与水、醇、醚、酮、不饱和烃、芳香烃等混溶，溶解能力强	低毒
二甲亚砜	dimethyl sulfoxide	DMSO	与水、甲醇、乙醇、乙二醇、甘油、乙醛、丙酮、乙酸乙酯、吡啶、芳香烃混溶	微毒，对眼有刺激性
吡啶	pyridine	Py	与水、醇、醚、石油醚、苯、油类混溶，能溶解多种有机物和无机物	低毒，对皮肤黏膜有刺激性